Naturalism and Its Alternatives

Naturalism and Its Alternatives in Scientific Methodologies:
Proceedings of the 2016 Conference on Alternatives to Methodological Naturalism

Published in the United States by Blyth Institute Press in Broken Arrow, Oklahoma

Library of Congress Control Number: 2016959657

ISBN: 978-1-944918-07-1

For author inquiries, please send email to info@blythinstitute.org. For more information, please see www.blythinstitute.org.

Information about the Alternatives to Methodological Naturalism series of conferences (including videos of the original conference presentations) can be found at www.am-nat.org.

Information about The Blyth Institute can be found at www.blythinstitute.org.

1st printing

Blyth Institute Press

Naturalism and Its Alternatives

in Scientific Methodologies

EDITED BY
JONATHAN BARTLETT
AND ERIC HOLLOWAY

Proceedings of the 2016 Conference on
Alternatives to Methodological Naturalism

This book is dedicated to the truths that exist in all the places where people tell us not to look.

"Theoretical pluralism" is better than "theoretical monism"

— Imre Lakatos, *The Methodology of Scientific Research Programmes*

It is a puzzling thing. The truth knocks on the door and you say, "Go away, I'm looking for the truth," and so it goes away.

— Robert M. Pirsig, *Zen and the Art of Motorcycle Maintenance*

Acknowledgements

This volume has been a joint effort between a number of collaborators. To begin with, we want to thank the Center for Evolutionary Informatics (CEI) for funding the conference upon which this book is based. Without their help, the conference would not have been nearly as successful.

We also want to thank Mario López for his assistance with the conference and numerous other details. Mario built a great website for the conference (am-nat.org), helped out with the conference itself, and has contributed to the conference and the book in numerous other ways.

Next, we want to thank Heather Zeiger for editing assistance. Heather has a unique ability to make papers clearer, more readable, and more technically correct (both in language and meaning) without sacrificing the voice of the author.

We also want to thank the many reviewers that assessed and critiqued papers for this volume. David Nemati, Abraham Lewis, Stephen Kunath, Iwan Sandjaja, Gary Prok, Vincent Torley, Kirk Durston, Cornelius Hunter, Jerry Bergman, and others all provided valuable insight and feedback that greatly improved the quality of these papers. In addition to the official reviewers, we know that there are many unofficial reviewers that each author used to improve their work before submission, and we are all grateful for the input of colleagues and friends who sharpen us each and every day.

Finally, we want to thank all of the contributors to this volume. Developing methodologies takes time and thought, and the payoffs, while they can be large, are usually distantly down the road. These are important conversations for the long run, but rarely do such conversations have immediate payoffs. We appreciate each and every one of you for taking the time and effort to research and write these papers. We also want to thank the people who presented at the conference but did not submit to this volume or submitted elsewhere. We appreciate having you as part of our conversation.

—Jonathan Bartlett and Eric Holloway

Contents

Introduction

JONATHAN BARTLETT AND ERIC HOLLOWAY

The Blyth Institute

1   Naturalism and Its Role in Science

Naturalism is the belief that every cause is a "natural" cause. In this sense, "natural" does not mean "existing in nature," but rather that the cause is of a material nature. That is, non-material causes, such as purpose, will, the soul, transcendental causes, supernatural beings, and God, are considered invalid and are not causally-effective entities. According to naturalism, all causation must eventually be reduced to the action of physics on material objects and forces (McMullin, 2011).

Naturalism is often divided into two different ideas, *philosophical* naturalism (also known as *metaphysical* naturalism) and *methodological* naturalism. Philosophical naturalism states that naturalism, as defined above, is the way that things really are. There is no God, there are no souls or spirits, and what you think of as your free will is merely an expression of your ignorance about the physics that forces your direction, or it is an outgrowth of the randomness of quantum causes. Methodological naturalism, on the other hand, doesn't pretend to know what exists or doesn't exist. Rather, methodological naturalism simply says that, when doing science, we should all behave *as if* naturalism were true. That is to say that, as part of its methodology, science should exclude all subjects that are not valid subjects within naturalism.

Thus, while philosophical naturalism is a metaphysical position (it describes the nature of the world and what does and does not exist), methodological naturalism is a procedural position. It describes the types of entities that can be considered and what types of cause and effect are valid within a framework (Forrest, 2000).

2   Science and Methodologies

Science is often defined and demarcated by its methodologies (Kojonen, 2016). The problem, however, is that such demarcation criterion often cuts in bad ways. That is, when a methodological demarcation is used to split science from non-science, it usually removes things from "science" ideas and areas of study that are intuitively connected with science. Alternatively, the methodological demarcation can actually include things that seem intuitively disconnected from science.

As an example, some people consider experimentation a unique feature of science. However, evolutionary biology often proceeds without experimentation, and, in fact, as the eminent evolutionary biologist Ernst Mayr declares, laws and experiments are often inappropriate techniques for evolutionary biology (Mayr, 2000).

Many people also use testability as a criterion for science. However, in such a case, string theory, at least in its current formulation, would considered non-science (Ellis and Silk, 2014). If this is the case, what should we do? Expel string theory from science departments? Make string theorists move to the philosophy department? If the answer is "nothing," then there is no reason to bother making demarcations at all.

So what *does* make something a science? Sich (2012) gives a much wider view of science that includes any sort of knowledge that can be obtained by demonstration. Here, demonstration does not mean that one has to physically *do* something to demonstrate it, but it can include logical demonstrations. The point is that science is knowledge that is obtained through connecting the dots, wherever and however those dots appear. The different sciences are then distinguished by their subject matter.

Under this rubric, most attempts to gain knowledge within a subject area can be classified as science. What differs is the subject matter. Different subject matters, then, have differing methods of attaining answers to questions. Truth in mathematics is demonstrated by appeals to formal proofs; truth in physics is demonstrated by appeals to experiment; truth in evolutionary biology is demonstrated by appeals to the fossil record, etc.

The methodologies employed in each case depend on the subject matter that is being investigated. Thus, if there was a valid way to demarcate science from non-science, it would be that the methodology employed (i.e., the mode of demonstration) properly matched the subject matter being investigated.

Under this view of science, the hurdle for investigating non-naturalistic causes would be the establishment of methodologies that are appropriate to the subjects being investigated. It is claimed by many proponents of methodological naturalism that methodological naturalism is required to gain reliable knowledge (Forrest, 2000).

Some contend that methodological naturalism is not a prerequisite for knowledge, but it is a prerequisite for science (Kojonen, 2016). However, under this view, the term "science" ceases to do any real work except group together a few subjects for no apparent reason. Even according to Kojonen (2016), each individual science has its own individual methodologies. Therefore, haphazardly grouping a few chosen subjects that have little in common into a bucket doesn't seem to actually achieve any particular goal, except perhaps to express a vague historical connection between a few subjects.

Whether or not methodological naturalism has ever had any usefulness in any particular science, prior to the 1800s it was primarily connected with physics, for which it is probably most appropriate. However, in the 1800s, the growth of philosophical naturalism enabled the introduction of methodological naturalism within sciences that had previously not included it. Darwin introduced methodological naturalism into biology; Freud introduced methodological naturalism into psychology; and Marx introduced methodological naturalism into economics (though it was presaged somewhat by Adam Smith's invisible hand) (Dilley, 2013; Futuyma, 1998). Prior to this, the reduction of these sciences to physics was far from a foregone conclusion. Though there are examples of individuals throughout history who believed that all of life is reducible to physics, it is primarily within this time period that this became a presupposed rule for *all* sciences, not just the physical sciences.

3 Naturalism's Methodological Takeover of Inquiry

One reason that naturalism became entrenched as a methodology is a failure to have competing methodologies. In the sciences that had significant methodological traditions of their own, such as mathematics and logic, naturalism has had minimal impact. That's not to say that there were not many naturalists within mathematics pursuing mathematics based on naturalism. David Hilbert, Alfred North Whitehead, and Bertrand Russell were all pursuing a mathematical naturalism, as exemplified by the *Principia Mathematica*. Because the *methodology* of mathematics did not depend on naturalism, the ideas of Cantor and Gödel were still able to thrive.[1]

However, no other subject had a strong enough existing methodology to stand in the path of naturalism. Therefore, as the naturalistic methodologies grew in both capability and coverage, more and more subject matters began to adopt them.[2]

---

[1]For those unaware of the relationship between infinity and theology as related to mathematics, see Drozdek (1995).

[2]On a side note, one might view theology as being the least likely area of inquiry to adopt a naturalistic viewpoint. However, I find it interesting that there is somewhat of a bifurcation in theology between those that adopt naturalism and those that do not. Historically, Charles Hodge (1872) introduced a methodology for systematic theology. This method, based on the inductive method from science (but using the Bible as the source of data), has been the foundation of theological inquiry in evangelical circles. While the methodology has many drawbacks (acknowledged by those both inside and outside of evangelicalism), it is interesting that evangelical theology that stems from Hodge's inductive methodology has been the least immune to the rise of naturalism. To get a sense for what theology looks like under naturalism, many modern theologians have turned to viewing theology as merely a unique form of social dynamics. That is, in this view, theology is not about God *per se*, but about human thought about God. Rather than trying to understand God, or who God is, theology outside of evangelicalism largely tries to naturalize the supernatural by making God the outgrowths of our own desires—personal, communal, and political. Even within evangelicalism, those who eschew Hodge often turn to naturalistic methodologies within theology.

It has been tacitly assumed that naturalism is the only methodology capable of producing results, or, for that matter, the only perspective capable of producing a methodology. Many, even among those who voice their opposition to methodological naturalism, fail to consider that there could be non-naturalistic methodologies. Therefore, it is understandable that in the decision between having a methodology and not having a methodology, most people who want a rigorous pursuit of truth choose to have a methodology. Those same people tend to choose naturalistic methodologies for lack of an alternative.

4 Stepping Outside of Naturalism

However, naturalism, both as a metaphysic and as a methodology, has been shown to have numerous weaknesses. Over the past few decades, methodological naturalism has had a consistent stream of scientists, philosophers, and other academics criticizing its position (Gordon, 2011; Plantinga, 1998). Even philosophical naturalists have sometimes seen the problems with adopting a rigid naturalistic stance, not the least of which is the ability to defend the theory of evolution in an objective manner (Boudry, Blancke, and Braeckman, 2010; Sarkar, 2011).

Likewise, in recent years, work has been done to show the value of non-naturalistic methodologies in various scientific and engineering enterprises. In 2000, the *Nature of Nature* conference at Baylor University brought together an array of eminent scientists and philosophers to discuss the merits and problems of naturalism in science (Gordon and Dembski, 2011). Several of those opposed to the rule of naturalism continued to meet and work together throughout the years.

The *2012 Conference on Engineering and Metaphysics* was held by The Blyth Institute to bring together some of this work and ask whether or not non-naturalistic methodologies can be put to *practical* use. The proceedings of this conference were published in the book *Engineering and the Ultimate: An Interdisciplinary Investigation of Order and Design in Nature and Craft*. This work included:

- How theological and philosophical frameworks can be utilized to improve contemporary architecture (Hall, 2014)

- How teleologically-driven approaches to science often result in improved results,

---

(See, for instance, Murphy (1999), who adopts "physicalism," which is naturalism for everything except God.) I would argue that while it is certainly possible to academically analyze subjects without having a definitive methodology, the simple act of *having* a methodology (whether good, bad, or otherwise) so assists the cohesiveness of academic enterprises that it allows deeper and firmer interactions among and between scholarly works so as to move a field forward. Thus, when faced with the absence of a methodology and the presence of one, even a bad methodology tends to win out over none at all, even when the methodology clashes with the subject matter as naturalism does with theology.

especially in biology (Halsmer, Gewecke, Gewecke, Roman, Todd, and Fitzgerald, 2014)

- How the a non-naturalistic view of creativity can improve models of cognition (Bartlett, 2014b)

- How software complexity can be better managed by use of non-naturalistic models (Bartlett, 2014a)

- How to detect and measure human capacity to perform algorithmically hard problems (Holloway, 2014)

- How to detect the impact of teleology on an outcome (Ewert, Dembski, and Marks, 2014)

- How a holistic view of self-replication can improve the investigation of origin-of-life scenarios (Mignea, 2014)

5 Developing Alternatives to Methodological Naturalism

These glimpses from the *Engineering and Metaphysics* conference gave promise to the possibility of non-naturalistic science. However, while these papers provided isolated cases where non-naturalism in science can work, none provided a template that could rightly be called a methodology. Therefore, another conference was scheduled with the goal of developing methodologies that were not based on naturalism within various sciences. This conference was the *2016 Conference on Alternatives to Methodological Naturalism*, and this present volume contains the proceedings of that conference. The goal was not necessarily to develop full-blown methodologies in any particular discipline, but to have an interdisciplinary conversation about where, why, and how alternatives to methodological naturalism might be able to work.

The present volume is divided into three parts. Part I discusses the problems with naturalism and demarcation in the sciences. These papers express the shortcomings of methodological naturalism generally, while also proposing general solutions based on the nature of inquiry itself.

Part II goes deeper into how alternatives to methodological naturalism might be built, including how to incorporate various aspects of non-naturalistic thinking into methodologies. This part includes papers on mathematics, teleology, analogy, theology, and other topics elucidating how these bits and pieces play into the development of knowledge in different subjects. It also includes potential drawbacks and issues that can arise in using non-naturalistic methodologies, and how these might be mitigated.

Finally, Part III details specific suggestions for specific fields. This part covers fields from economics to computer science, and shows both large and small ways that alternatives to methodological naturalism might be incorporated into the subject.

The *Nature of Nature* conference established the need for other approaches to science. The *Engineering and Metaphysics* conference showed that non-naturalistic approaches to knowledge can indeed be fruitful. Now, the *Alternatives to Methodological Naturalism* conference starts the discussion on how this might be implemented in practice on a larger scale.

Read and think. Then decide if these ideas can be applied in some way to your own field and to your own questions, and whether these ideas can improve your own methodologies.

References

Bartlett, J. 2014a. Measuring software complexity using the halting problem. In J. Bartlett, D. Halsmer, and M.R. Hall (editors), *Engineering and the Ultimate: An Interdisciplinary Investigation of Order and Design in Nature and Craft*, pp. 123–130, Blyth Institute Press, Broken Arrow, OK.

Bartlett, J. 2014b. Using Turing oracles in cognitive models of problem-solving. In J. Bartlett, D. Halsmer, and M.R. Hall (editors), *Engineering and the Ultimate: An Interdisciplinary Investigation of Order and Design in Nature and Craft*, pp. 99–122, Blyth Institute Press, Broken Arrow, OK.

Boudry, M., Blancke, S., and Braeckman, J. 2010. How not to attack Intelligent Design Creationism: Philosophical misconceptions about methodological naturalism. *Foundations of Science* 15(3):227–244.

Dilley, S. 2013. The evolution of methodological naturalism in the Origin of Species. *HOPOS: The Journal of the International Society for the History of Philosophy of Science* 3(1):20–58. http://www.jstor.org/stable/10.1086/667897

Drozdek, A. 1995. Beyond infinity: Augustine and Cantor. *Laval Théologique et Philosophique* 51(1):127–140.

Ellis, G. and Silk, J. 2014. Scientific method: Defend the integrity of physics. *Nature* 516:321–323.

Ewert, W., Dembski, W.A., and Marks, R.J. 2014. Algorithmic specified complexity. In J. Bartlett, D. Halsmer, and M.R. Hall (editors), *Engineering and the Ultimate: An Interdisciplinary Investigation of Order and Design in Nature and Craft*, pp. 131–152, Blyth Institute Press, Broken Arrow, OK.

Forrest, B. 2000. Methodological naturalism and philosophical naturalism: Clarifying the connection. *Philo: A Journal of Philosophy* 3(2):7–29.

Futuyma, D. 1998. *Evolutionary Biology*. Sinauer, 3rd edition.

Gordon, B.L. 2011. The rise of naturalism and its problematic role in science and culture. In B.L. Gordon and W.A. Dembski (editors), *The Nature of Nature: Examining the Role of Naturalism in Science*, pp. 3–61, ISI Books.

Gordon, B.L. and Dembski, W.A. (editors). 2011. *The Nature of Nature: Examining the Role of Naturalism in Science*. ISI Books.

Hall, M.R. 2014. Truth, beauty, and the reflection of god: John ruskin's *Seven Lamps of Architecture* and *The Stones of Venice* as palimpsests for contemporary architecture. In J. Bartlett, D. Halsmer, and M.R. Hall (editors), *Engineering and the Ultimate: An Interdisciplinary Investigation of Order and Design in Nature and Craft*, pp. 65–96, Blyth Institute Press, Broken Arrow, OK.

Halsmer, D., Gewecke, M., Gewecke, R., Roman, N., Todd, T., and Fitzgerald, J. 2014. Reversible universe: Implications of affordance-based reverse engineering of complex natural systems. In J. Bartlett, D. Halsmer, and M.R. Hall (editors), *Engineering and the Ultimate: An Interdisciplinary Investigation of Order and Design in Nature and Craft*, pp. 11–38, Blyth Institute Press, Broken Arrow, OK.

Hodge, C. 1872. *Systematic Theology*. T. Nelson and Sons, London.

Holloway, E. 2014. Complex specified information (csi) collecting. In J. Bartlett, D. Halsmer, and M.R. Hall (editors), *Engineering and the Ultimate: An Interdisciplinary Investigation of Order and Design in Nature and Craft*, pp. 153–166, Blyth Institute Press, Broken Arrow, OK.

Kojonen, E.V.R. 2016. Methodological naturalism and the truth seeking objection. *International Journal for Philosophy of Religion* .

Mayr, E. 2000. Darwin's influence on modern thought. *Scientific American* 283(1):66–71.

McMullin, E. 2011. Varieties of methodological naturalism. In B.L. Gordon and W.A. Dembski (editors), *The Nature of Nature: Examining the Role of Naturalism in Science*, pp. 82–94, ISI Books.

Mignea, A. 2014. Developing insights into the design of the simplest self-replicator and its complexity. In J. Bartlett, D. Halsmer, and M.R. Hall (editors), *Engineering and the Ultimate: An Interdisciplinary Investigation of Order and Design in Nature and Craft*, pp. 169–220, Blyth Institute Press, Broken Arrow, OK.

Murphy, N. 1999. Physicalism without reductionism: Towards a scientifically, philosophically, and theologically sound portrait of human nature. *Zygon* 34(4):551–570.

Plantinga, A. 1998. Methodological naturalism? *Origins and Design* 18(1–2).

Sarkar, S. 2011. The science question in Intelligent Design. *Synthese* 178(2):291–305.

Sich, A. 2012. The independence and proper roles of engineering and metaphysics in support of an integrated understanding of God's creation. In J. Bartlett,

D. Halsmer, and M.R. Hall (editors), *Engineering and the Ultimate: An Interdisciplinary Investigation of Order and Design in Nature and Craft*, pp. 39–59, Blyth Institute Press, Broken Arrow, OK.

Part I

Naturalism and the Demarcation Problem

Naturalism is generally considered foundational for the definition of science, but is that a valid demarcation criterion? If naturalism is the divide between science and non-science, then how much does this matter? These papers focus on how naturalism relates to the demarcation problem, and how to demarcate science in absence of naturalism.

Philosophical Shortcomings of Methodological Naturalism and the Path Forward

Jonathan Bartlett

The Blyth Institute

Abstract

Methodological naturalism, when used to enforce an exclusive view of scientific investigation, is based on three problematic streams of philosophy: mechanical philosophy, positivistic epistemology, and divine incomprehensibility. Each of these philosophies has inherent flaws that prevent them from being usable across the entirety of causal relationships that science attempts to investigate. However, even in the face of such criticisms, methodological naturalism as a methodology does have some positive features that should be retained even if methodological naturalism itself is not.

1 Defining Methodological Naturalism

Methodological naturalism has had many different proponents and definitions over the years. Trying to sort these out can be rather difficult, as they each have their own idea of what methodological naturalism is, why it is important, and what it means for the scientific enterprise. Proponents of methodological naturalism have included those who think that non-naturalistic explanations are within the bounds of science but are not very fruitful as well as those who think that such explanations should be *a priori* removed from scientific inquiry.

Most claims about methodological naturalism fall into the *a priori* exclusionism camp. Likewise, the *a priori* exclusionism view is the most important because it can be used, and has been used, as a rule for one party to exclude the investigations of another party from a field of science.

This paper will focus on the exclusionist view for two reasons. First of all, since we are looking at the issue from a philosophical standpoint, the question of whether or not some individuals find naturalism personally helpful or not is not very interesting. What people today find helpful says little about ontology or epistemology in the broader sense. Second, strict exclusionism is used as a rule to exclude the findings of others. That is, the whole point of strict exclusionism is not for someone to say what they themselves are doing or why they themselves have chosen a particular methodology, but to exclude from conversation others who have chosen differently. In order to rationally justify such an exclusion, one needs to establish a firm basis for doing so. If I prefer to study tree frogs over velociraptors, this is a choice that I may have reasons for, but that I do not need to justify to anyone other than myself. If I want to prevent other people from studying or publishing findings about velociraptors, then I need to have strong philosophical grounds for doing so, and such grounds must be open to scrutiny by others. Likewise, if someone wishes to focus on naturalistic causes and methods in their own study, there is no reason why anyone else should care. However, if someone wishes to prevent someone else from focusing on non-naturalistic causes and methods, they need to have strong grounds for such an action.

The most authoritative statement on the question of methodological naturalism comes from the National Science Teachers Association (NSTA), which has been used by other authoritative groups such as the National Academy of Sciences (NAS). The most commonly-referenced version of this statement appears in the 1998 book *Teaching about Evolution and the Nature of Science* published by the NAS. In this book, the NAS republished the NSTA's statement as follows:

> Science is a method of explaining the natural world. It assumes the universe operates according to regularities and that through systematic investigation we can understand these regularities. The methodology of science emphasizes the logical testing of alternate explanations of natural phenomena against empirical data. Because science is limited to explaining the natural world by means of natural processes, it cannot use supernatural causation in its explanations. Similarly, science is precluded from making statements about supernatural forces because these are outside its provenance. Science has increased our knowledge because of this insistence on the search for natural causes.

> (Working Group on Teaching Evolution, National Academy of Sciences, 1998, pg. 124)

The NSTA has updated their statement over the years, though it is the same in spirit. Their new statement says,

> Science is a method of testing natural explanations for natural objects and events. Phenomena that can be observed or measured are amenable

to scientific investigation. Science also is based on the observation that the universe operates according to regularities that can be discovered and understood through scientific investigations. Explanations that are not consistent with empirical evidence or that cannot be tested empirically are not a part of science. As a result, explanations of natural phenomena that are not derived from evidence but from myths, personal beliefs, religious values, philosophical axioms, and superstitions are not scientific. Furthermore, because science is limited to explaining natural phenomena through testing based on the use of empirical evidence, it cannot provide religious or ultimate explanations.

(National Science Teachers Association, 2013)

Both of these statements are intended to be normative for scientific practice, have been used as the basis for the development of science standards, and have been used to justify the exclusion of other forms of inquiry from science (NGSS Lead States, 2013; SC Education Oversight Committee, 2014; Scharmann, 2005; Lerner, 2000; Katskee, 2006). These statements, because they were issued by normative agencies (the NAS and the NSTA) and because of their history of use for the exclusion of inquiry, will be used as the focus for this paper. The paper will focus especially on the 1998 statement because of its widespread distribution through the National Academy of Sciences. Quotes of the statement will refer to the 1998 statement.

2 Locating the Intellectual Sources of the Statement

In recent years, many scientists have disregarded philosophy as a valid source of knowledge. In fact, in the NSTA's updated statement above, the NSTA explicitly rejects philosophical axioms as a source of scientific knowledge. The problem with rejecting philosophy as a source of knowledge is that it also undercuts the foundations of science itself. Most scientific investigations operate based on the *principle of sufficient reason* and the *identity of indiscernibles* (Bartlett, 2014a). These principles come to science from philosophy, and, as such, indicate that nearly all scientific knowledge actually originates from philosophical axioms. If a person tests a substance to determine that its properties match that of hydrogen and concludes that it *is* hydrogen, then that person is going beyond the empirical data and using the identity of indiscernibles as a philosophical axiom to reason from the empirical data to the conclusion that the substance is hydrogen.

In fact, as has been pointed out by many philosophers of science, even so-called empirical facts are model-based and therefore based on philosophical axioms (Polanyi, 1946; Hanson, 1958; Quine, 1968; Feyerabend, 1993; Polkinghorne, 1998). As the present paper will show, many of these models themselves are based on personal beliefs and what the NSTA would certainly classify as "myths."

However, the present point is that scientists as a group, especially scientific organizations representing scientists, seem to be wholly ignorant of the role that philosophy plays in their reasoning. As such, it is very difficult to derive *from scientists* an explicit explanation of the underlying philosophy for what they do and how they describe what they do. This does not mean that they do not employ philosophy, but merely that they are unaware of what that philosophy is, where it comes from, and what the known limitations are with a given philosophy.

Every philosophy brings with it knee-jerk prejudices. The goal of philosophy is not to eliminate these, but to bring them under examination. When a person thinks that they don't use philosophy, what it usually means is that they are engaging in the knee-jerk prejudices of a philosophy that they are unaware of, and therefore are unable to critically examine. Therefore, the goal of assessing the philosophical underpinnings of these statements of methodological naturalism is to bring them fully to the surface so they can be examined in the light of day and not just held as mere prejudices.

The goal of this paper is to show the original streams of philosophy that undergird methodological naturalism (as exemplified by the NSTA statements), and then offer a critique of their ability to serve as a philosophical foundation for science.

The philosophy behind a particular scientific methodology would not matter so much if that methodology operated within an isolated scientific pursuit. Then it could be argued quite reasonably that the methodological restrictions emanate not from some particular universal philosophy, but by the nature of the subject matter itself. However, methodological naturalism is not just a claim about a specific science, but about all sciences. While science originated in physics, it now encompasses a diverse set of fields including biology, anthropology, psychology, cosmology, and evolution, each with their own individual subject matters. Methodological naturalism claims that each of these, because they are sciences, must entail the same methodological restrictions as physics. Additionally, two of the subjects mentioned—cosmology and evolution—are *totalizing* subjects. That is, the truth of their findings depends on taking into account *all* available causes; one cannot simply pick out an individual, restricted subject matter and focus on a subset of causes and have the results be valid. Cosmology, for instance, covers the entire history of the universe. If there are causes in play anywhere in the universe that are not covered by methodological naturalism, then those working in cosmology may be missing important causes simply by the methodological restrictions imposed from the outside. Likewise, this is true for evolution. If there are causes in play anywhere in the history of life that are not covered by methodological naturalism, then those working in evolution may be missing important causes necessary to discovering the truth about the evolution of life on Earth. Again, since these are totalizing subjects, such restrictions cannot be seen as emanating from the subject matters themselves since the subjects include every available cause in the subject's history and not just those derived within a specific context (i.e., the context of a lab or experiment).

Therefore, we can also say that because methodological naturalism is used in science no matter what the subject matter is and because it is used in totalizing sciences as well, methodological naturalism must be based not on the subject matter but on a governing philosophy of nature and epistemology. As such, it is imperative to uncover and analyze the philosophies behind methodological naturalism.

3   The Three Streams of Philosophy

This paper argues that methodological naturalism has its source in three streams of thought: mechanical philosophy, positivist epistemology, and theological incomprehensibility. These streams provide synergy to each other with each justifying the underlying presuppositions of the other. The result of these streams working together is methodological naturalism in science as exemplified by the NSTA statements.

The first stream is mechanical philosophy. Mechanical philosophy is a philosophy of nature originating in the 1600s. It has taken several forms (which will be explored), but in essence, the goal is to view nature as machine-like. It can be understood as a reaction to the scholastic philosophy of nature popular in the Middle Ages. The philosophers of the 1600s viewed the explanations given by the scholastic philosophers as essentially non-explanations (Slowik, 2014). The scholastics viewed the world in terms of substantial forms and qualities. To the mechanical philosophers, these forms and qualities seemed to require explanations themselves. To say that something is "dry" because it has more of the quality "dryness" seems to not actually explain dryness. Likewise, saying that something is circular because its form has a circular nature is not actually saying anything about why it possesses that shape.

Therefore, mechanical philosophy says that fuller explanations of the world can be made by treating the world as a machine and asking what sorts of mechanisms create the forms and qualities we see in nature.

In the nineteenth century, mechanical philosophy gave birth to positivism and its cousin, pragmatism, as epistemologies. Positivism is the idea that any statement that cannot be understood in terms of matter and motion (i.e., the fundamental terms of mechanical philosophy) is not only wrong, but entirely non-sensical. For instance, the word "love" is only meaningful if someone can give an operational definition of love—for example, "Love means that someone will hug me often." In positivism, all terms must eventually be definable in operational (i.e., mechanical) terms to have any valid meaning. Positivism denied the reality of most spiritual beings, forces, and ideas by stating that since their ultimate meaning cannot be given in efficient terms, they cannot be knowable entities.

Pragmatism is similar to positivism but takes a slightly different tack. Positivism aims at real knowledge and limits real knowledge to things that can be known in efficient terms. Pragmatism, on the other hand, does not seek real knowledge, but instead it seeks ideas with a "cash value." Pragmatism does not care if *any* of its terms or relationships have any basis in reality to them at all. Pragmatism only cares whether or not certain considerations of reality prove useful. There are many ways that pragmatism can be made amenable to spiritual qualities. Even its founder thought that pragmatism could encompass spiritual ideas, saying, "If theological ideas prove to have a value for concrete life, they will be true, for pragmatism, in the sense of being good for so much" (James, 1907). For the interested reader, an extensive and thoughtful treatment of the pragmatism of such ideas in society can be found in Niebuhr (1952).

Despite its roots as a neutral player in religious and non-religious thought, American pragmatism over the years has largely coalesced with positivism. The "cash value" of positivism is usually construed as operational results—i.e., an idea is good if it provides predictive value of the future. Therefore, pragmatism is commonly used to exclude theological propositions on similar grounds as positivism does, but without making any metaphysical statement about the reality or unreality of such.

In short, positivism and pragmatism are very similar in practice with positivism viewing non-operational statements as "meaningless" and pragmatism viewing them as "worthless."

Given that most people are religiously inclined (Hackett and Grim, 2012), and, in fact, that religious inclination may be a biological feature of humans (Boyer, 2008; Barrett, 2011), it would be surprising for these philosophies by themselves to have a large-scale impact on human thinking. However, a theological movement originating in the nineteenth century paved the way for positivistic ideas to establish themselves among the theologically-inclined.

The spiritual realm has always been viewed as a mystery to humans. Perhaps it is touchable, but it is largely seen as incomprehensible. Likewise, the Christian understanding of God asserts that God's ways are not always understandable to humans. For example, Isaiah 55:8 says that God tells man, "For my thoughts are not your thoughts, neither are your ways my ways, declares the LORD. For as the heavens are higher than the earth, so are my ways higher than your ways and my thoughts than your thoughts."

In Christian thought, this has historically been taken to be a partial assertion. That is, to some extent God's thoughts and ways can be knowable, but ultimately God's thoughts and ways are above us. However, this idea that God's ways are above ours has, throughout history, often been used as an "escape hatch" for those engaging in apologetics. That is, if someone says that "God would not have allowed X in the world," an apologist may respond, "God's ways are higher than our ways, so it is possible that God has a really good reason for allowing X in the world." The problem is that if any positive statement about theology can avoid disputation by simply stating that God's ways are beyond ours, then it is difficult to see how any theological dispute may be resolved.

Starting in the nineteenth century, this led to the doctrine of divine mystery to be transmogrified into the doctrine of divine incomprehensibility. That is, not only is God mysterious, but there is no objective way to know or say anything about God. Therefore, nothing that we say about God can constitute knowledge. Originally, this began merely as a question of what humans can know about *God* but eventually was applied to all forms of spiritual knowledge. Today, while the doctrine of divine incomprehensibility is rarely stated as such, it can be seen in practice by the way that society treats theological knowledge as inherently personal and non-arguable. Theological propositions are personal because, though they might be true, there is no way to adjudicate differences of opinion. Therefore, they are treated as personal opinions and not the subject of public discourse. Because of divine incomprehensibility, they stand as both non-refutable and unknowable. Likewise, each individual's theological opinions are treated with equal weight because they are all equally non-refutable and unknowable.

These three philosophical streams—mechanical philosophy, positivism/pragmatism, and divine incomprehensibility—even though each does not necessarily entail the other, are mutually reinforcing. Mechanical philosophy provides an ontology of the world (or at least the natural world) that consists entirely of matter and motion. Positivism and pragmatism restrict epistemology to only quantities that are relevant to mechanical philosophy. From the other direction, divine incomprehensibility removes theological ideas from the realm of rational discourse, leaving only the terms that are used in mechanical philosophy.

Positivism views theological ideas as meaningless. Pragmatism views theological ideas as unhelpful. Mechanical philosophy views theological ideas as unreal. Divine incomprehensibility views theological ideas as real but unknowable. What is left over from this subtraction is, essentially, mechanical philosophy. While divine incomprehensibility usually presumes (contra mechanical philosophy) that there is a real spiritual aspect to the world, the fact that no definitive statements about God or the spiritual realm can be made makes it sufficiently innocuous for those operating within mechanical philosophy to countenance. Divine incomprehensibility also moves pragmatism closer to mechanical philosophy. As we noted earlier, it is possible for pragmatic theological statements to be developed. However, divine incomprehensibility discourages such claims by *a priori* determining them to have no cash value. While pragmatism makes no definitive claims about the nature of the world, it lines up very cleanly with mechanical philosophy.

So, while these philosophies are largely independent of each other, they exist together in a comfortable, mutually-reinforcing space. In other words, the knee-jerk prejudices of each of these philosophies line up and reinforce each other.

Figure 2.1: The Mutually-Reinforcing Nature of the Three
Streams of Methodological Naturalism

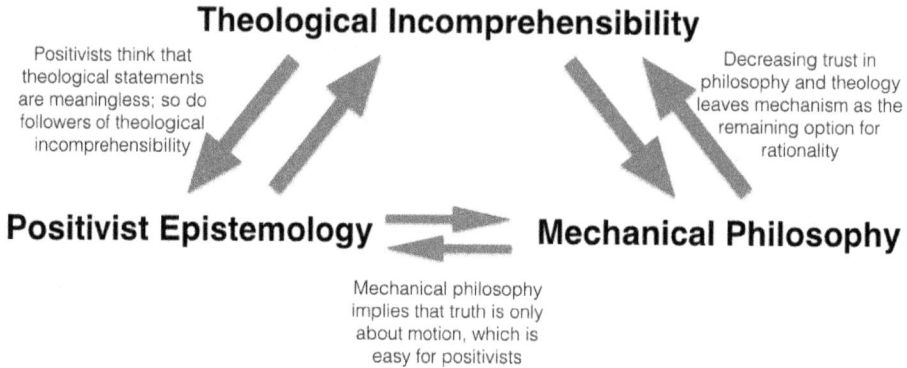

Theological Incomprehensibility

Positivists think that
theological statements
are meaningless; so do
followers of theological
incomprehensibility

Decreasing trust in
philosophy and theology
leaves mechanism as the
remaining option for
rationality

Positivist Epistemology   **Mechanical Philosophy**

Mechanical philosophy
implies that truth is only
about motion, which is
easy for positivists

4   Tracing These Views in the NSTA Statements

What we have established so far is the mutually-reinforcing nature of mechanical philosophy, positivist/pragmatist epistemology, and divine incomprehensibility. The question then is how does this trio add up to methodological naturalism? Each of these philosophies can quite easily be picked out from the NSTA statements themselves.

The starting assumption of the statement is that science "assumes the universe operates according to regularities and that through systematic investigation we can understand these regularities." First of all, it assumes that the universe operates according to regularities. It could have said that *some parts* of the universe operate according to regularities. Instead, it made the blanket statement that the universe operates according to regularities. This is quite consistent with mechanical philosophy.

Next, note that it doesn't say that we can understand *some* of these regularities or a *subset* of these regularities, but just that we can understand these regularities. Thus, epistemology is likewise limited to the contents of mechanical philosophy. Further, it states that science is based on testing against empirical data (i.e., matter and motion), which is from the positivist/pragmatist view of reality.

Finally, the statement deals with religious explanations. It says that "science is limited to explaining the natural world by means of natural processes, it cannot use supernatural causation in its explanations." However, this statement is never justified. What is it about the nature of the world that prevents its regularities from having a supernatural cause? There are only two possible explanations: either this is merely a subject-matter distinction, or it is because the major proponents of this view believe mechanical philosophy to be a true statement about the world.

Let's start by considering that this is merely a subject-matter distinction. That is, science doesn't include supernatural causes because it is too busy focusing on natural causes. Non-natural causes may be interesting to some people but that is not what these individuals have chosen to study. In such a view, the subject matter is the operating demarcation and methodologies flow naturally from the nature of the subject matter under investigation (Sich, 2014). However, the way that the NSTA statement is constructed, the NSTA is talking about the universe of inquiry into causation, not a subset of it.

Additionally, subject-matter restrictions never surface as absolute prohibitions. For instance, if I were to say that Renaissance studies were limited to the time period of the fourteenth century to the seventeenth century, there would be no opprobrium attached to a paper discussing thirteenth century precursors to the Renaissance, or its influence in the eighteenth century and beyond, or even about the overlap between the Renaissance and its precursors and after-effects. There would be no legal cases trying to use the courts to prohibit the introduction of such ideas into Renaissance studies.

Were the restriction merely one of subject matter, it is difficult to see how or why anyone would object to someone blurring the lines, finding overlaps, or identifying other kinds of integrational activities between science and non-science causes. Additionally, for the subject-matter restriction to make sense, it would need to provide some means of distinguishing between a cause that is within the realm of science and one that is not. Because such a distinction is not provided, it appears that the writers of the statement don't believe that any such cause exists.

Furthermore, the statement says, "science is precluded from making statements about supernatural forces because these are outside its provenance." Again, without a means of distinguishing between science and supernatural forces, it seems that the author thinks that all regularities and everything that can be rigorously investigated has nothing, and can have nothing, to do with the supernatural. This only makes sense on the grounds of divine incomprehensibility. If the universe contained supernatural causes, and a scientist rigorously investigated a cause that he did not yet know was supernatural, what would the NSTA statement have the scientist do? Destroy his data? Burn his books? Apologize for studying the wrong subject? No, it seems that the statement simply assumes that any rigorous investigation into the knowable world would only lead to naturalistic causes, indicating that those are either the only ones available or, assuming divine incomprehensibility, are the only ones that are intelligible.

Finally, the statement says, "Science has increased our knowledge because of this insistence on the search for natural causes." In this world of interdisciplinary cooperation, it seems strange that science alone benefits from a provincial view of knowledge. The only way to make sense of this provincial view is if its proponents believe it is the only game in town.

While the statement does not make explicit reference to mechanical philosophy, positivism/pragmatism, or divine incomprehensibility, the influence of these philosophies can be readily apprehended. Therefore, to see if the foundation of this reasoning is correct, we need to look at each of these philosophies in turn to see if they are both correct in and of themselves and if they are helpful to the progress of science when strictly employed. I will make no attempt to answer the question as to whether these philosophies are helpful in individual cases (as I believe that they indeed have been), only to whether a *strict* adherence to them, and to the rules laid out in the NSTA statements, would have a negative impact on the progress of science.

5   Mechanical Philosophy

The ontology that sits behind all of this is mechanical philosophy—the idea that all of nature, including our minds, is the product of machine-like forces.

5.1   Mechanical Philosophy in the 1600s

Mechanical philosophy arose as a contender against scholastic Aristotelian views of nature that defined things in terms of substances and qualities. The proponents of mechanical philosophy felt that Aristotelian explanations did not really explain, and therefore sought to explain, nature in a way that gave deeper causal explanations. While Aristotelian philosophy was a philosophy of wholes, mechanical philosophy focused on breaking things down into parts and figuring out how the parts themselves worked. Aristotelian philosophy focused on beings and their natures and qualities, while mechanical philosophy focused on how the parts of beings worked together to make such natures and qualities work.

As such, this does not quite make an ontological distinction. What made mechanical philosophy so different was its relentless goal of re-envisioning everything in terms of material and efficient causes and the disregard of any form of non-material causation.

Whether true or not, this seems straightforward on the surface. The problem, however, comes in laying down what one really means by a "material cause." The problem of defining material causes has plagued mechanical philosophy from the beginning. In fact, some of the things that today are viewed as the high achievements of mechanical philosophy were, in their day, in ardent opposition to mechanical philosophy.

Mechanical philosophy in the 1600s, for instance, while not completely uniform, generally held the following views:

Atomic Corpuscularism This is the view that, at its core, nature is based on the geometry of tiny, impermeable particles and their interactions. Some viewed these particles as rigid, though some thought they had some amount of flexibility at their core. Differently shaped atoms led to different types of substances and interactions.

Localism This is the view that change can only happen locally. That is, for one particle to influence another, it must be touching that particle. Action at a distance is considered a spiritual mode of causation.

Gradualism This is the view that there are no instantaneous changes in nature. Everything must go through a smooth series of steps. In fact, this criteria is one of the reasons why some viewed atoms as being flexible—a fully rigid atom, when bounced against another, would have to instantaneously change direction. By allowing atoms to flex, it preserved both atomism and gradualism.

Passivism This is the idea that changes happen *to* atoms. That is, atoms just "are" and their geometry affects how they interact with each other. Atoms are basically passive and do not actively affect the world around them.

These views formed the basis of mechanical philosophy in the 1600s. Note that this does not just describe the physics of the 1600s, but is tied to the mechanical philosophy that was being developed at that time. Atoms were passive, not because of experiment, but because mechanical philosophy only allowed geometry to influence the actions of physics. The world was made of atoms because the higher-level beings described by the scholastics were too spiritual. Causation was local because in order to posit action at a distance, one had to suppose that there was some quality of matter beyond its geometry—again, harkening to the spiritualized scholastic view of the world.

Surprisingly, the major advance in physics in the 1600s didn't come out of mechanical philosophy, but out of Newtonian physics, which broke with nearly every aspect of mechanical philosophy. Much of Newton's work was based on gravity, which was much more scholastic than mechanical.

First of all, gravity is an inner power, a quality of an object—not the result of any geometry of that object. As a quality, gravity harkens back to the scholastic ways of thinking, which considered nature in terms of substances, powers, natures, and qualities. Second, gravity exerts action across a distance. Prior to Newton, the ability to exert action across a distance was reserved for spiritual beings and forces. Finally, with gravity, objects are active instead of passive players in physics. Objects themselves, by their nature, exert forces on the things around them. They are not passive as required by mechanical philosophy.

Thus, Newtonian physics, while held up by many today as the triumph of mechanism over religious views of the world, was actually the opposite in its day. Newton was the one who violated methodological naturalism by imputing qualities to nature that were previously reserved for spiritual beings and forces. In fact, this was one of the most common criticisms of Newtonian physics at the time it was presented.

Gottfried Liebniz was one such critic:

It is also a *supernatural thing* that bodies should attract one another at a distance without any intermediate means and that a body should move around without receding in the tangent, though nothing hinders it from so receding. For *these effects cannot be explained by the nature of things.*

(Liebniz, 1716, emphasis added)

As you can see, Liebniz, co-inventor of Calculus, criticized Newton for proposing a supernatural force and even re-emphasized that these forces were non-natural. Some, such as Einstein, have proposed more localized versions of Newtonian mechanics that didn't succumb to Liebniz's criticisms. However, the question is, what should Newton have done with his own investigations? Should he have abandoned them since they didn't conform to the naturalistic expectations of his day? Should he have waited for the time when he could make a naturalistic version of them? If not, then on what basis should someone today allow naturalistic expectations of nature to impede on an investigation? As we see with Newton, what was viewed as non-mechanistic in one era is viewed as the epitome of mechanism in another.

5.2   Mechanical Philosophy in the 1800s

As noted earlier, trying to pin down mechanical philosophy is somewhat difficult. After the success of Newtonian physics, a new mechanical philosophy had to be invented which included the Newtonian perspective. Therefore, in the 1800s, the new mechanical philosophy focused on these points:

Causal Closure of Nature This held that there is nothing besides nature itself that leads to the outcomes of nature. Nature is composed of fixed laws, and once the laws are known and the present state of the world is known, everything is known about the causes that will generate the next series of effects.

Determinism Because of the causal closure of nature, if all of the forces acting in nature are known, then the effects can be known as well (in theory, though no one believes this would be true in practice). Since the current effects are the causes of the next set of effects, then the cause/effect cycle can be theoretically followed as far into the future as one wishes. Thus, all future effects can be, in principle, determined from the current state of the world.

Matter-focused Because one of the purposes of mechanical philosophy is to remove spiritual modes of causation from nature, mechanical philosophy is fundamentally matter-focused.

This new mechanical philosophy is best summarized by the mathematician Pierre Laplace:

We may regard the present state of the universe as the effect of the past and the cause of the future. An intellect which at any given moment knew all of the forces that animate nature and the mutual positions of the beings that compose it, if this intellect were vast enough to submit the data to analysis, could condense into a single formula the movement of the greatest bodies of the universe and that of the smallest atom; for such an intellect nothing could be uncertain and the future just like the past would be present before its eyes.

(Laplace, 1814)

This new mechanical philosophy seemed strong and formidable. However, at the beginning of the twentieth century, new breakthroughs in physics destroyed this view of mechanical philosophy as well. The rise of quantum mechanics showed that all physical descriptions are partial and incomplete. Not only is quantum mechanics itself incomplete, but experiments have shown that randomness (i.e., the inability to fully describe causation in terms of its present state) is fundamental to physics. This destroys both the causal closure of nature as well as determinism.

Likewise, quantum mechanics brought about a shift in thinking away from matter-focused physics. The wave/particle duality of quantum physics means that a matter-focused philosophy of nature was inherently deficient. In fact, in some branches of quantum mechanics (such as the Copenhagen school), mind is itself considered a fundamental entity.

Finally, the one aspect of mechanical philosophy of the 1600s that Newton did not contradict, gradualism, was in fact overturned by quantum mechanics, which says that causation can and does happen in leaps.

Therefore, for a second time, the advances in physics were not in line with the expectations of mechanical philosophy and naturalism. While naturalism has certainly successfully retrofitted itself to make room for each of these advances in physics, it is difficult to make the case that naturalism in any way helped push physics forward, except perhaps as a means of codifying the assumptions of the present age that needed to be challenged. It certainly does not seem that science should be limiting itself based on a particular era's view of what counts as a naturalistic explanation.

5.3   Modern Mechanical Philosophy

With every iteration, mechanical philosophy has grown quite a bit weaker as a statement about reality. The mechanical philosophy of the 1600s made fairly definite statements about what it thought "mechanism" did and did not include. While this philosophy turned out to be largely mistaken on its view of the world, we can at least appreciate it as an intelligible description of how they thought the world worked.

In the mechanical philosophy of the 1800s, the focus was not on a definitive description of the world, but on a general model of how causes and effects interacted in any suitable description of the world. In other words, they skipped over what the specific laws of physics should be and just focused on the relationship between the laws and the things governed by those laws.

In the modern incarnation of mechanical philosophy, a mechanism is considered anything that is:

- A distinct phenomena

- Composed of parts

- Parts that have a structure

- Parts that are in some form of a causal relationship with each other

While there is probably nothing wrong *per se* with this form of mechanical philosophy, it is difficult to see what, if anything, it has to say for a scientist investigating a phenomenon. Even someone investigating the paranormal (ghosts, goblins, etc.) would be able to frame their investigations in terms of this definition of mechanism. In fact, it is difficult to imagine any description of events that would not be consistent with this form of "mechanism." As such, if this is all that is meant by "natural phenomena," it is difficult to envision what it is that the NSTA statements were trying to exclude.

Therefore, if the NSTA statement is attempting to make a description of something with the term "natural phenomena" it is difficult to see what it is that they mean. Mechanical philosophy is usually what that term is thought to include, but, as we have seen, mechanical philosophy has repeatedly been shown to be false, or, in the modern incarnation, it is fairly meaningless as a way of demarcating phenomena. In fact, as the history of mechanical philosophy has shown, "natural phenomena" seems to only signify the prejudices of the current age, which holds science back, rather than a template for moving science forward.

One other possible meaning of "natural phenomena" in the modern era, which has a more definitive structure, is computationalism, which is described in Stephen Wolfram's Principle of Computational Equivalence (Wolfram, 2002) and other similar principles, such as the Tractable Cognition Thesis (van Rooij, 2008). The idea behind these principles is that the limits of nature are the same as the limits of computation. For this philosophy, a "natural phenomenon" is one that is computable with a Turing-like machine.

Like the previous versions of mechanical philosophy, this view likely serves to encapsulate the prejudices of the current era, which need to be overcome, rather than actually describing the real state of the world. While a full criticism of this view will not be presented in this paper, the reader can read more on this subject in Bartlett (2014b) and in Bartlett (2016).

6 Positivism and Pragmatism

Even though mechanical philosophy is untenable as an ontological basis for methodological naturalism, this does not necessarily invalidate the positivist/pragmatist epistemology. For instance, it is possible that even if specific theories of the world do not need to rely on mechanical philosophy, our knowledge of them may be limited to the operational ways in which these theories affect our experience.

As noted earlier, depending on what someone's goals are, it is possible to formulate pragmatism in a manner outside the realm of positivism. It is *often* aligned with the positivistic view of things, but this is not a necessary equivalence. Additionally, if one's goal in science is to discover reality, pragmatism is not a suitable epistemology. Pragmatism is often used as a "compromise" or "placeholder" epistemology, where if the ontological status of a statement is in question, one can appeal to the pragmatic value of the statement even if the ontological status is largely doubted.

As such, the only view of epistemology that would be foundational for a strict exclusionist view of methodological naturalism would be positivism. Therefore, the question becomes, can positivism form a *normative* basis of scientific epistemology for the entire world of inquiry into causation in the universe?

It is possible that in a limited frame of reference, positivism may be a worthwhile epistemology just like naturalism may be a worthwhile ontology for some aspects of the world. However, science, as practiced today, does not just include physics, but also biology, psychology, evolution, cosmology, and others. Cosmology and evolution, being total views of cosmic history, likewise require total views of causation in order to form an accurate picture of their subject. Therefore, for an epistemology to be normative for science generally, and not a specific science, it must be universally applicable. As such, we will show that positivism is an inadequate total epistemology.

The fundamental problem with positivism is that it is self-refuting. Positivism claims that all statements that are meaningful must only be defined operationally (i.e., in terms of empirically-verifiable results) or must be true *a priori* (i.e., mathematical statements). However, the claim of positivism itself cannot be defined in terms of empirically-verifiable results, and it is not true *a priori*. Therefore, if positivism were true, it would refute itself as meaningless. That alone should exclude it from being a total epistemology.

However, for reasons that are incomprehensible to the present author, some do not think that self-refutation (a common problem in skeptical epistemologies) invalidates a philosophical position. Therefore, we will present additional problems with positivistic views.

The first problem with positivism is Quine's ontological relativity. Quine points out that all statements, including empirical ones, are part of a larger web of belief structures. Therefore, in order to make empirical statements, we must have in place prior belief structures that are not empirical that allow us to make our measurements. As Hanson points out, all empirical measurements are model-based. Since our measurements are founded upon these models and belief structures, the measurements can only be as true as the underlying models and belief structures upon which they are based. Therefore, for empirical results to be meaningful, the non-empirical models upon which empirical results are based must also be meaningful.

The second problem with positivism is Gödel's incompleteness, which states that there are true, knowable facts that are not verifiable from a fixed set of axioms. The only allowed exceptions to empirically verifiable truths allowed by positivism are *a priori* truths such as mathematics. However, Gödel's incompleteness introduces a new set of mathematical truths that are *based* on mathematics, but are not verifiable axiomatically from mathematics and are not verifiable empirically. At least some of these truths are knowable. Therefore, there are real, important truths that are not verifiable in the ways positivism lays out.

Third, there are similar problems with non-mathematical knowledge. As Plantinga points out, life is full of real but unverifiable truths. For instance, one cannot verify the consciousness of other minds. The existence of such is something that we all presume, but it cannot be verified in any external fashion. One may argue that we have a biological (and therefore *a priori*) foundation for such beliefs, but that would mean we also have a biological basis for believing in God and other spiritual beings. Therefore, if positivism includes biological foundations in the set of *a priori* knowledge that can be used in science, then one could not exclude supernatural explanations on that basis.

However, positivism and pragmatism are not completely empty as epistemologies. One quality that they emphasize, and is worth keeping, is the connectedness of truths to each other. Positivism, by connecting all knowledge to operational action, forces knowledge to at least be connected to *something*. This does in fact help eliminate meaningless truths that are completely unconnected to reality. The problem, though, is that assuming everything must be connected in a certain way to a particular set of truths overly restricts knowledge and meaning.

7   Divine Incomprehensibility

As mentioned previously, divine incomprehensibility is an approach to theological ideas that says theological knowledge is super-rational. This term is sometimes used for the notion that there are *aspects* of God that are not comprehensible to us (Sproul, 2014). However, here we are referring to the movement that gained headwinds in the nineteenth century and goes beyond this saying that everything about God (except possibly God's existence) is unknowable. Since God's ways are above our ways, and God's nature is above our nature, there is no way for us to have any real knowledge about God or any supernatural subject. Divine incomprehensibility is technically limited to knowledge about God, but it is usually, as will be shown, then applied to all theological and non-material knowledge.

There is a question as to what theology motivated the early development of empiricism in science. Voluntarism, the idea that God created the world freely, in the way that He chose, and without compulsion, is often viewed as the foundation of empiricism (Oakley, 1961; Henry, 2009). Since the choice to create was a free choice of God, *what* he created was also a free choice and therefore could not be deduced from *a priori* knowledge. Instead, we must go out into the world and study it to learn what it was God chose to create. Harrison (2002) has contended that a weaker form of divine incomprehensibility stood at the heart of the early empiricists. Since the scientists could not know God's thoughts, they must therefore discover what the world was like through empiricism. At that time in history, while their *a priori* knowledge of God's intentions were viewed skeptically, the *ability to learn* God's intentions through empirical study was not.

In the nineteenth century, however, the doctrine of divine incomprehensibility had started to take hold in academic circles, and its influence paved the way for modern ideas about methodological naturalism.

This is can be seen, for instance, in the modifications to Darwin's *Origin of Species* throughout its various editions. Darwin continually revised *Origin* in response to criticisms of his work. These revisions can give insight to the way in which Darwin justified his methods and inferences. Dilley (2013) has catalogued the changes to *Origin* that deal with methodological naturalism (though not under that name) and its justifications.

As Dilley points out, in the first three editions, nothing like methodological naturalism appeared. It isn't that Darwin did not believe in methodological naturalism (Gillespie (1979) notes that he believed in it from early in his career), but rather that either Darwin didn't feel he needed to include it as a justification, or, as Dilley suggests, he believed that the intellectual climate at the time would not accept it as a justification.

In the sixth edition, he appealed implicitly to divine incomprehensibility as the justification for excluding supernatural explanations in biology. The following was added to the sixth edition (emphasis mine):

> He who believes that some ancient form was transformed suddenly through an internal force or tendency into, for instance, one furnished with wings, will be almost compelled to assume, in opposition to all analogy, that many individuals varied simultaneously. It cannot be denied that such abrupt and great changes of structure are widely different from those which most species apparently have undergone. He will further be compelled to believe that many structures beautifully adapted to all the other parts of the same creature and to the surrounding conditions, have been suddenly produced; and of such complex and wonderful co-adaptations, *he will not be able to assign a shadow of an explanation*. He will be forced to admit that these great and sudden transformations have left no trace

of their action on the embryo. To admit all this is, as it seems to me, to *enter into the realms of miracle, and to leave those of Science.*

(Darwin, 1861, pg. 205)

In other words, if one asserts that a miracle occurs, then no shadow of explanation can be determined in principle. The implicit reason is that God's ways are incomprehensible to us.

Additionally, owing to the more expanded view of divine incomprehensibility, this is not just the case for God, but for *any* non-naturalistic explanation. Darwin has applied this doctrine to not only apply to miracles, but also to internal non-mechanical forces of organisms. Darwin classes those sorts of mechanisms in *the same category* as that of miracle. No justification is given, except for the suggestion that "no shadow of explanation" can be determined if this is the case.

This mode of justification has persisted into modern times, as can be seen in several writings about methodological naturalism. However, this idea of divine incomprehensibility is usually implicitly, rather than explicitly, made. The explicit argument is typically made regarding God's omnipotence rather than His incomprehensibility. The form this modern argument takes is that since God can do anything, anything that occurs is logically consistent with God having done it capriciously. As an example, Forrest (2000) says,

> Introducing supernatural explanations into science would destroy its explanatory force since it would be required to incorporate as an operational principle the premise that literally anything which is logically possible can become an actuality, despite any and all scientific laws; the stability of science would consequently be destroyed.

Boudry, Blancke, and Braeckman (2010) point out that this and similar statements are not actually problematic, as it merely restates the definition of logical possibility. However, if one includes the doctrine of divine incomprehensibility as an unstated foundation, the statement becomes much more convincing. If God's actions were comprehensible, then God's omnipotence would not pose the problem described—finding the comprehensibility of God's actions would be just like finding the comprehensibility of anything else. It is only if we presume God to be incomprehensible that this poses a real problem for science.

Additionally, Forrest's claim is not just about God, but about *all* supernatural explanations. Only under the more expansive view of divine incomprehensibility does the ineffability of God extend to every other aspect of the supernatural. Otherwise, finding the comprehensibility of the supernatural would not be radically different than finding the comprehensibility of anything else. One needs only to establish a methodology for reducing the logically possible to the actually possible.

As Boudry et al. point out, this problem is not unique to Forrest, but extends to most modern defenses of methodological naturalism in science.

In summary, the empiricism of the seventeenth century was established by theological voluntarism, which only rejected our ability to know God's thoughts in an *a priori* manner. However, in the nineteenth century, this transformed into a total skepticism of our ability to comprehend or obtain knowledge of anything about God or even anything supernatural or non-natural. This is the implicit justification that makes sense of modern claims about theological knowledge in defenses of methodological naturalism.

It is interesting that the justification for methodological naturalism relies not on empirical facts, but rather on theological questions of what can and cannot be determined through empiricism. If the goal is to exclude theological thought in the development of science, the irony is that methodological naturalism does not exclude it but instead takes sides on theological questions.

Not only that, but Darwin's rejection of final causes made biological knowledge incomprehensible. If one follows Darwin, then *any* discussion of final causes in biology is non-scientific. This led to generations of biologists avoiding any kind of teleological language in biology. In the early twentieth century, biologists would avoid saying things like, "the animal made a nest *in order to* lay eggs in it," as that was too teleological. Instead, biologists would often phrase it as "the animal made a nest *and then* laid eggs in it," which reduced the actions to non-teleological, mechanical descriptions.

Pittendrigh (1958) and Mayr (1961) re-established teleological language in biology through the use of the concept of teleonomy. Even so, it is interesting that despite the fact that teleological behaviors are the primary ways in which organisms function, it took many decades to find a loophole around methodological naturalism in order to talk about teleology in a way that is compatible with methodological naturalism. Thus, rather than aiding in the investigation of the subject matter, such restrictions were a setback to biological investigation. One could also argue that Pittendrigh and Mayr did not successfully naturalize teleology in biology, and that, in fact, biology generally operates from teleological assumptions couched in naturalistic language and terminology (Bedau, 1991; Fodor and Piattelli-Palmarini, 2010).[1] In either case, it is evident that naturalism did not arise from the study of the subject matter, but instead was the result of a certain flow of theological reasoning.

In fact, theological reasoning has been foundational to many aspects of science. Experimental biology was developed by Francesco Redi to prove that organisms only originate by God after their kind and not by spontaneous generation.[2] Mendel (1865)

---

[1] As a further example of this, systems biology shows the productivity of a biology that holds the purpose of the system as methodologically equal, or even prior to, the understanding of its components (Lander, 2004; Noble, 2006; Snoke, 2014). While many participants of systems biology would consider their approach naturalistic, the purpose-first methodology is actual counter to this understanding.

[2] Redi (1688) states, "I shall express my belief that the Earth, after having brought forth the first plants and animals at the beginning by order of the Supreme and Omnipotent Creator, has never since produced any kinds of plants or animals, either perfect or imperfect; and everything which we

developed genetics to show that variation has fixed limits (i.e., within created kinds).[3] A Catholic priest developed the Big Bang Theory , which aligned Thomistic theology and Genesis with cosmology.[4]

As such, it is difficult to maintain, given that several of the overarching theories of modern science were developed in a theological context, that somehow theology cannot in principle add knowledge to inquiry. Instead, it seems that theology has actually been fairly successful at providing frameworks for inquiry.

8   Re-Establishing Non-Naturalistic Knowledge

In summary, nearly every part of the NSTA statement on methodological naturalism has been shown to rest on faulty philosophy or theology. While methodological naturalism may make sense in certain contexts, the use of it as a total stricture on every type of scientific inquiry, especially totalizing inquiries, is not philosophically sound.

However, one should not discount the benefits that methodological naturalism has had for science, We can use this to learn how to include non-naturalistic knowledge into science. First of all, methodological boundaries make inquiry easier by limiting the scope of inquiry. Every person who has written a dissertation knows the importance of limiting the scope of inquiry in order to make definitive statements. Such limits are required for successful extrapolation from the evidence. Second, methodological naturalism contained within it (via positivism) a system of warrant and justification.

Thus, methodological naturalism reduced the scope of inquiry enough to make many inquiries manageable, and it provided a system of justification that one could

---

know in past or present times that she has produced, came solely from the true seeds of the plants and animals themselves, which thus, through means of their own, preserve their species." Thus, he began his experiments to prove the truth of Genesis and started the field of experimental biology.

[3]The last two paragraphs of Mendel's paper indicate this. He says, "Gärtner, by the results of these transformation experiments, was led to oppose the opinion of those naturalists who dispute the stability of plant species and believe in a continuous evolution of vegetation. He perceives in the complete transformation of one species into another an indubitable proof that species are fixed with limits beyond which they cannot change. Although this opinion cannot be unconditionally accepted we find on the other hand in Gärtner's experiments a noteworthy confirmation of that supposition regarding variability of cultivated plants which has already been expressed." Carl Friedrich von Gärtner (1772–1850) was a prominent creationist at that time, and here Mendel is describing how his system of genetics aligns well with Gärtner's results.

[4]Many people think that Lemaître entirely separated his science from his theology. However, the evidence is that Lemaître privately aimed at arriving at a cosmology that integrated his Thomistic theology with science. In public, Lemaître made statements such as "As far as I can see, such a theory remains entirely outside any metaphysical or religious question" (Holder, 2012, pg. 50). However, his unpublished work showed that he viewed the Big Bang as validating Genesis (Holder, 2012, pg. 49). Additionally, there are reports that in private conversations early in the development of his theory that he viewed the Big Bang as a reconciliation of Thomism with science (Peratt, 1988, pg. 196).

use to externally validate claims. This system does not work perfectly and is not always used. However, having a framework in place gives everyone who practices science a common reference point for thinking about and analyzing claims. I will define a *methodological framework* as a combination of limiting inquiry and a system of justification. As a methodological framework, methodological naturalism limits itself to objects exhibiting mechanical behavior and uses positivism as its system of justification.

In order to successfully do scientific work outside of methodological naturalism, two things are needed. First, new methodological frameworks are needed for different subject matters and types of inquiry. Second, a means of integration between methodological frameworks is needed for integrating knowledge between disciplines and studies. As an example, Rakover (2016) introduces a methodological framework aimed at psychology. It is a beginning step, but it includes limits to the subject matter, a system of justification, and some rudiments on how it can be integrated with other types of inquiry.

Science itself, as well as the modern conception of methodological naturalism, came out of philosophical and theological reasoning. As Sich (2014) points out, science cannot neither establish its own principles nor validate its own ability to guide the investigator to truth. Philosophy and theology are required for these things. Therefore, we should look toward philosophy and theology to further establish new methodological frameworks for investigation, to establish the principles by which they operate, and to validate their abilities to guide the investigator to truth.

References

Barrett, J.L. 2011. *Cognitive Science, Religion, and Theology: From Human Minds to Divine Minds*. Templeton Press.

Bartlett, J. 2014a. Introduction. In J. Bartlett, D. Halsmer, and M.R. Hall (editors), *Engineering and the Ultimate: An Interdisciplinary Investigation of Order and Design in Nature and Craft*, pp. 1–8, Blyth Institue Press, Broken Arrow, OK.

Bartlett, J. 2014b. Using turing oracles in cognitive models of problem-solving. In J. Bartlett, D. Halsmer, and M.R. Hall (editors), *Engineering and the Ultimate: An Interdisciplinary Investigation of Order and Design in Nature and Craft*, pp. 99–122, Blyth Institue Press, Broken Arrow, OK.

Bartlett, J. 2016. Describable but not predictable: Mathematical modeling and non-naturalistic causation. In J. Bartlett and E. Holloway (editors), *Naturalism and Its Alternatives in Scientific Methodologies*, Blyth Institue Press, Broken Arrow, OK.

Bedau, M. 1991. Can biological teleology be naturalized? *The Journal of Philosophy* 88(11).

Boudry, M., Blancke, S., and Braeckman, J. 2010. How not to attack intelligent design creationism: Philosophical misconceptions about methodological naturalism. *Foundations of Science* 15(3):227–244.

Boyer, P. 2008. Religion: Bound to believe? *Nature* 455:1038–1039.

Darwin, C. 1861. *On the Origin of Species by Means of Natural Selection, or the Preservation of Favoured Races in the Struggle for Life*. John Murray, London, 6 edition.

Dilley, S. 2013. The evolution of methodological naturalism in the origin of species. *HOPOS: The Journal of the International Society for the History of Philosophy of Science* 3(1):20–58.
http://www.jstor.org/stable/10.1086/667897

Feyerabend, P. 1993. *Against Method*. Verso.

Fodor, J. and Piattelli-Palmarini, M. 2010. *What Darwin Got Wrong*. Farrar, Straus and Giroux.

Forrest, B. 2000. Methodological naturalism and philosophical naturalism: Clarifying the connection. *Philo* 3(2):7–29.

Gillespie, N. 1979. *Charles Darwin and the Problem of Creation.* University of Chicago Press, Chicago.

Hackett, C. and Grim, B.J. 2012. The global religious landscape. *The Pew Forum on Religion and Public Life* .
http://www.pewforum.org/2012/12/18/global-religious-landscape-exec/

Hanson, N. 1958. *Patterns of Discovery.* Cambridge University Press.

Harrison, P. 2002. Voluntarism and early modern science. *History of Science* 40.

Henry, J. 2009. Voluntarism at the origins of modern science. *History of Science* 47.

Holder, R.D. 2012. Georges lemaître and fred hoyle: Contrasting characters in science and religion. In R.D. Holder and S. Mitton (editors), *Georges Lemaître: Life, Science, and Legacy*, pp. 39–55, Springer.

James, W. 1907. *Pragmatism: A New Name for an Old Way of Thinking.* Harvard University Press.

Katskee, R.B. 2006. Why it mattered to dover that intelligent design isn't science. *First Amendment Law Review* 5:112–146.

Lander, A.D. 2004. A calculus of purpose. *PLoS ONE* 2(6).

Laplace, P. 1814. A philosophical essay on probabilities .

Lerner, L.S. 2000. *Good Science, Bad Science: Teaching Evolution in the States.* Thomas B. Fordham Foundation.
https://edex.s3-us-west-2.amazonaws.com/publication/pdfs/lerner_7.pdf

Liebniz, G. 1716. Liebniz's fourth reply to dr. clarke .
http://www.earlymoderntexts.com/assets/pdfs/leibniz1715_1.pdf

Mayr, E. 1961. Cause and effect in biology. *Science* 134(3489):1501–1506.

Mendel, G. 1865. Versuche über pflanzenhybriden. *Verhandlungen des naturforschenden Vereines in Brünn* 4.
http://mendelweb.org/Mendel.plain.html

National Science Teachers Association. 2013. Nsta position statement on the teaching of evolution .
http://www.nsta.org/about/positions/evolution.aspx

NGSS Lead States. 2013. Appendix h: Understanding the scientific enterprise: The nature of science in the next generation science standards. In *Next Generation Science Standards: For States, By States*, pp. 430–436, National Academies Press, Washington, D.C.

Niebuhr, R. 1952. *The Irony of American History.* Scribner.

Noble, D. 2006. *The Music of Life: Biology Beyond the Genes.* Oxford University Press.

Oakley, F. 1961. Christian theology and the newtonian science: The rise of the concept of laws of nature. *Church History* 30:433–457.

Peratt, A.L. 1988. Dean of the plasma dissidents. *The World and I* pp. 190–197. http://plasmauniverse.info/downloads/DeanOfPlasma.pdf

Pittendrigh, C. 1958. Adaptation, natural selection, and behavior. In A. Roe and G.G. Simpson (editors), *Behavior and Evolution*, pp. 390–416.

Polanyi, M. 1946. *Science, Faith, and Society.* Cambridge University Press.

Polkinghorne, J. 1998. Critical realism in science and religion. In *Belief in God in an Age of Science*, pp. 101–124, Yale University Press.

Quine, W.V. 1968. Ontological relativity. *The Journal of Philosophy* 65(7):185–212.

Rakover, S. 2016. Methodological dualism and a multi-explanation framework: An approach needed for understanding behavior. In J. Bartlett and E. Holloway (editors), *Naturalism and Its Alternatives in Scientific Methodologies*, Blyth Institue Press, Broken Arrow.

Redi, F. 1688. *Experiments on the Generation of Insects.* Translated by Mab Bigelow in 1909. https://archive.org/download/experimentsongen00redi/

SC Education Oversight Committee. 2014. Special panel of the State Board of Education and EOC regarding High School Biology Standard H.B.5. (july 29, 2014) . http://www.eoc.sc.gov/Home/Cyclical%20View%20of%20Science% 20Standards/July%2029,%202014%20Special%20Panel%20Meeting.pdf

Scharmann, L.C. 2005. A proactive strategy for teaching evolution. *The American Biology Teacher* 67(1).

Sich, A. 2014. The independence and proper roles of metaphysics in support of an integrated understanding of god's creation. In J. Bartlett, D. Halsmer, and M.R. Hall (editors), *Engineering and the Ultimate: An Interdisciplinary Investigation of Order and Design in Nature and Craft*, pp. 39–62, Blyth Institue Press, Broken Arrow, OK.

Slowik, E. 2014. Descartes' physics. In E.N. Zalta (editor), *The Stanford Encyclopedia of Philosophy*.
https://plato.stanford.edu/archives/sum2014/entries/descartes-physics/

Snoke, D. 2014. Systems biology as a research program for intelligent design. *BIO-Complexity* 2014(3).

Sproul, R.C. 2014. Divine incomprehensibility. *Tabletalk Magazine* .
http://www.ligonier.org/learn/articles/divine-incomprehensibility/

van Rooij, I. 2008. The tractable cognition thesis. *Cognitive Science: A Multidisciplinary Journal* 32(6).
http://staff.science.uva.nl/~szymanik/papers/TractableCognition.pdf

Wolfram, S. 2002. *A New Kind of Science*. Wolfram Media.

Working Group on Teaching Evolution, National Academy of Sciences. 1998. *Teaching About Evolution and the Nature of Science*. National Academy Press, Washington, DC.
http://www.nap.edu/read/5787/chapter/1

3 || # Methodological Naturalism, Methodological Theism, and Regularism

TOM GILSON

ThinkingChristian.net

My practice as a scientist is atheistic. That is to say, when I set up an experiment I assume that no god, angel or devil is going to interfere with its course; and this assumption has been justified by such success as I have achieved in my professional career. I should therefore be intellectually dishonest if I were not also atheistic in the affairs of the world.
—J.B.S. Haldane, Geneticist, 1892-1964.

Abstract

Scientists typically justify methodological naturalism on grounds that the supernatural (or extra-natural) is not testable, that admitting the supernatural (or extra-natural) into science would undermine scientific methodology and reasoning processes, and that methodological naturalism has been demonstrated to be effective. These admitted virtues of methodological naturalism are strongly associated, however, with unscientific metaphysical assumptions which tend to dominate scientific thinking even if they do not follow necessarily from methodological naturalism's assumptions. For that reason a metaphysically neutral alternative is called for, one that retains methodological naturalism's virtues while discarding its associated unscientific assumptions. Regularism, defined merely (and intentionally quite simply) as "the methodological expectation of reliable regularity of cause and effect in nature," fits these criteria, and is recommended as a superior alternative to methodological naturalism.[1]

---

[1]This paper was originally published by Touchstone, and is published here by special permission.

1   Introduction

A clearer, more concise statement of the functional view reigning among scientists today could hardly be found. God—or at least God's interference—is excluded from the lab, and it appears that there is good reason. Since science is the study of the natural order and its success rests on its assumption that the natural order is both natural and orderly, any supernatural forces meddling with it could only throw it into disarray.

The principle is illustrated nicely in a Sidney Harris cartoon. Two men in lab coats are standing at a blackboard with abundant mathematical terms to the left, more math to the right, and between them the statement, "Then a miracle occurs." One scientist points at the words and says, "I think you should be more explicit here in step two." We all laugh knowingly, even those of us who believe miracles actually can happen.

Science gets nowhere if it inserts miracles as explanations for natural processes. Maybe nature isn't all there is, but science must function as if it were so. Let the world disagree over gods, angels, and demons. For methodological purposes, scientists must assume naturalism. That is, they assume nothing exists (or nothing is going on, at least) except what happens with matter and energy interacting in time and space according to the regularities that we call natural law. Hence *methodological naturalism*, as this view is generally known. As Michigan State University philosopher of science Robert Pennock writes,

> Science operates by empirical principles of observational testing; hypotheses must be confirmed or disconfirmed by reference to empirical data. One supports a hypothesis by showing that consequences obtain which would follow if what is hypothesized were so in fact ...Supernatural theories...can give no guidance about what follows or does not follow from their supernatural components.

(Pennock, 2000).

Not only that, but (as Pennock goes on to say) admitting the supernatural into science would undermine scientific rationality. "Then a miracle occurs" is not merely unscientific. As a principle for understanding nature's regular functioning, it's also irrational.

Naturalism has more than methodological connections, however. There is also *philosophical naturalism*, which holds not only that scientists should act *as if* naturalism is true, but that it *really* is true. In other words, there is nothing in all reality except matter, energy, time, space, and natural law. It's a thoroughly atheistic perspective, and it is the view held by most atheists today, especially the so-called New Atheists. Let's set that aside for a moment, though. We ought to be able to agree that whatever a scientist believes on evenings and weekends, when she walks into the lab, she's a practicing atheist. It only makes sense, right?

Wrong. Judeo-Christian theism leads to exactly the same methodological expectations. We could just as accurately speak of methodological theism as a basis for science.

Thankfully, I'm making this suggestion from a safe place here. I'm quite sure that if I were to propose methodological theism in certain academic settings, the response would be just as predictable (and release as much uncontained energy) as an explosive chemical reaction in Haldane's lab: "Theism?! You're nuts! You can't import your religious beliefs into science! God has no place here!" My answer (if I could speak it amid all the uproar in the room) would be that I'm not really suggesting we try to bring theism either into the lab or the classroom. I'm leading toward something else instead. "Methodological theism" is just a step toward a different conclusion. Like a scaffold next to a wall, its usefulness lasts only while the thing is being built. Sure, it contains a widely unacceptable bias; I'm only using it provisionally to expose a different bias that also needs to be set aside.

To get to that point, however, I really need to show how theism can do the same methodological work that naturalism is expected to do.[2]

2   Theism and the Regularity of Nature

Science requires that nature be orderly and predictable. Theism does too. This is no mere post hoc Jewish or Christian accommodation to the success of the scientific method. It's well established in the very core of Christian theism for at least three reasons.

2.1   Human Moral Responsibility

First, theism expects that humans will be morally responsible agents. Theism holds that God gave us free will, the responsibility to choose either right or wrong, and that we are accountable for our moral choices. Moral responsibility is inextricably associated with being able to predict the results of our actions. I could illustrate this with myriad examples, but one should be enough: Does a mother have a moral responsibility to feed her child nutritious food? Undoubtedly. Suppose that on some days for some children broccoli was the great nutrition source we know it to be, and on other days (or for other children the same day) its effect became unpredictably, unknowably deadly. Suppose that meanwhile chocolate-frosted donuts were (again, unpredictably, unknowably) the food with the greatest nutritional benefit. Could a mother be held accountable for her child's death from broccoli? Could she be considered praiseworthy for not loading her children's plates with donuts? No. She

---

[2]The theism of which I speak, by the way, is Judeo-Christian theism as historically understood by the great majority of adherents. It may or may not apply to Muslim understandings of Allah. To specify "Judeo-Christian theism" every time would be tiresome, so I'll simply use "theism," with the understanding that I'm speaking of a certain theistic view throughout.

would be utterly helpless, morally speaking. She would have no moral responsibility for any such choices.

2.2 Human Learning

Second, theism expects that humans should be able to learn from experience, and that this is what God wants for us. "You reap what you sow," says the Christian Scriptures. Suppose we planted barley and reaped wheat one year, oats the next, and walnut trees the year after that. What would we learn about the effects of planting barley? Would we even call it barley seed? Would we know enough even to give it a name? Or what if irrigation or fertilizers sometimes (unpredictably, unknowably) killed the whole crop? We would know nothing about agriculture. "Fertilizer" wouldn't be just a foreign concept; it would forever be a blank spot in our minds. We would be as babes (or city-dwellers!) in our knowledge of how to feed a family, much less a community.

Regularity of cause and effect is important for science, but before there was science, it was crucial for other purposes as well. Men and women learned, through experience, where babies came from. Does God have no interest in that? I could multiply examples, but the point is made. Theism requires orderly natural processes so that we can learn.

2.3 Divine Communication

Third, theism holds that God has the ability and (at least sometimes) the intent to communicate understandably with humans. This, too, requires a generally orderly overall environment. We can borrow an engineering term, signal-to-noise ratio, to explain the reason. Consider a casual conversation in a quiet sitting room; then think about the same conversation in a crowded restaurant with hard walls and concrete floors—very live acoustics, in other words. The volume with which you speak to communicate successfully in one setting is totally inadequate for the other on account of the relatively higher noise.

In engineering terms, "noise" isn't always about sound. It refers to any functionally random energy that carries no useful information. (Background chatter in a restaurant isn't precisely random, but it is close enough for our purposes here.) For any signal to be expressed successfully, you always need a high enough signal-to-noise ratio to be able to pick out what is really signal as opposed to what is not.

Haldane imagines miracles interfering in the lab so often that we could not do science. If that was the case—if God's interference were both routine and unpredictable—miracles wouldn't be miracles; they would be noise instead. If axe heads floated on their own from time to time, the story in 2 Kings 6:1–6 wouldn't have been interesting enough to include in the Bible. If lepers spontaneously healed every once in a while, Jesus' touch would have communicated little. If men rose randomly from the dead, Jesus' resurrection would mean nothing.

For God to communicate through miracles, miracles must be truly exceptional events, which is exactly as theism holds them to be. Haldane's specter of angels or demons floating over the lab to muck things up is a fantasy of his own imagining. It has nothing whatsoever to do with Christian theism. The occasional, infrequent miracle does nothing to upset the scientific method. Scientists routinely encounter anomalous data, and they have a routine for what to do: either re-run the experiment, cast out the data point as a statistical outlier, or otherwise pick up and start over again. This is not unusual practice in science.

2.4   Theism and the Moral Basis of Science

There's another virtue required in science—honesty. Theism has been talking about that for a long time. It would be challenging to find a compelling reason for truth-telling in naturalism, which on its own implies no ethical requirements. Naturalists generally speak of doing what works best for various outcomes: human flourishing, the fulfillment of desire, or maximizing human good—including the progress of science. But, these are human add-ons to the completely amoral and impersonal naturalistic substrate. Theism, by contrast, sees honesty as a fundamental virtue reflecting the very nature of reality in a very personal God.

3   Unacceptable Metaphysical Biases

In a way, methodological theism makes just as much sense as methodological naturalism. It does the same thing for science that methodological naturalism does: it provides an expectation of orderliness and predictability. Bias remains a problem, however. Science ought to not spread that kind of metaphysical assumption all around the lab or, especially, in the classroom. For that reason, I am not proposing methodological theism as a universal principle for scientists to subscribe to. (Theists can, but they need not suggest or prescribe to it as a general rule for other scientists to follow.) Metaphysical baggage doesn't belong in science. I think all scientists should be able to agree with that. *This is exactly why methodological naturalism is also a poor choice for a principle of science.* Naturalism is, after all, a metaphysical position just as much as theism is. Inserting "methodological" before the word "naturalism" doesn't take the metaphysical position away from "naturalism" any more than it does with "theism," which we all recognize as unacceptably biased. Recall what Haldane (1934) said, "I should therefore be intellectually dishonest if I were not also atheistic in the affairs of the world."

Lawrence Krauss, professor of physics at Arizona State University, observed that Haldane

> understood that science is by necessity an atheistic discipline. As Haldane so aptly described it, one cannot proceed with the process of scientific discovery if one assumes a "god, angel, or devil" will interfere with one's experiments. God is, of necessity, irrelevant in science.
>
> Faced with the remarkable success of science to explain the workings of the physical world, many, indeed probably most, scientists understandably react as Haldane did. Namely, they extrapolate the atheism of science to a more general atheism.
>
> While such a leap may not be unimpeachable it is certainly rational...Though the scientific process may be compatible with the vague idea of some relaxed deity who merely established the universe and let it proceed from there, it is in fact rationally incompatible with the detailed tenets of most of the world's organized religions.

(Krauss, 2009)

Perhaps Krauss is unaware of the false dichotomy he sets up here. He sees only two options: either God is an interfering God, in which case science would be impossible, or God "merely established the universe and let it proceed from there." A third option—the one that accurately describes both Jewish and Christian doctrine in spite of his claim otherwise—is that God established the natural order to work regularly almost all of the time but with extremely rare exceptions. There's no need to assume naturalism of any sort in science.

4   Regularism Instead

All that science requires is that nature operate regularly. That's all methodological naturalism provides, conceptually, for science. Science does *not* require favoring one metaphysical viewpoint over another, provided that both are sufficiently orderly and regular. (Theism may provide better reasons to *expect* nature to act in a regular fashion, but that's a separate discussion). So neither methodological naturalism nor methodological theism serves very well as a universal principle for practicing scientists to follow. We need something metaphysically neutral for the job instead. We probably want something that works both in the lab and in everyday life so that a Haldane, a Krauss, or anyone including Christian scientists could live by the same principle with their white coats on or off. We need something that speaks merely of nature's regular operation.

And there's hardly anything simpler than that very word, "regular". So I conclude here by proposing we drop all the "methodological" business—all the metaphysical trappings, too—and realize that science operates like all of life on an expectation of natural regularity. If we must have an "ism" to describe it, let's use *regularism*. It's almost disappointingly simple. There's hardly anything to be added about it, even here at the point in the article where I introduce the term. As a view of reality it is extremely sparse. It can accommodate theism or naturalism equally well. It says what needs to be said. It adds nothing metaphysically extraneous, nothing unscientific, and nothing that needs to be put on or taken off at the laboratory door. It is a superior alternative to the metaphysically loaded term methodological naturalism. And, it is a better way to think about how we do science.

References

Haldane, J.B.S. 1934. *Faith and Fact*. Watts.

Krauss, L. 2009. God and science don't mix. *Wall Street Journal* .
http://www.wsj.com/articles/SB124597314928257169

Pennock, R.T. 2000. *Tower of Babel: The Evidence Against the New Creationism*. MIT Press, Cambridge, Massachusetts.

4

Design as a Criterion of Demarcation

Mario A. López

Organización Internacional para el Avance Científico del Diseño Inteligente

Abstract

Methodological naturalism, though inexplicit in the denial of purpose, operates exclusively under the tenets of *ontological naturalism*[1] and, therefore, proceeds only by way of the empirical and naturalistic. A more neutral epistemology is less presumptive and would allow science to flourish without the strictures of such a philosophical commitment. The task of divorcing science from methodological naturalism requires the abandonment of the idea that the structure of knowledge, or justified belief, requires no epistemic foundation (Neurath, 1959) and that inferential justification possess a uniquely superior epistemic status in the sciences than that which is non-inferentially known. As I see it, the persistent problem of science, and thus the criterion of demarcation that undergirds it, is two-fold. First, it is assumed that only inferential knowledge is genuinely justified, and second, that theories must be, at the very least, theoretically falsifiable.

In this paper, I intend to provide a criterion of demarcating science that is practical and heuristically useful to spur scientific progress. My proposition does not presuppose the causal powers of chance and necessity. Instead, it forces the scientist to appreciate the ontological characteristics of nature and to leave the question of causation completely open, thereby, avoiding the pitfalls that ontological naturalism, and its faithful ally, methodological naturalism, habitually impose on science.

1 Introduction

The history of science is replete with ideas of what science should be and how science should operate; however, delineating what science *is* has proven to be a difficult

---

[1]I will use ontological and metaphysical naturalism interchangeably.

task. Although definitions have not generally interrupted what happens in a lab, for knowledge of the world to progress, we need a clear distinction between what it means to actually do science and what it means to merely pretending to do science. We need to have an understanding of what we are looking for and a methodology of how to look for it. The standard (and dare I say, ambiguous) pronouncement is that science is devoted to solving problems, and it does so by using the observable physical world as the basis for solving them. This, in turn, increases our understanding of the world itself.

    This is all well and good, but surely we don't believe that the physical world is our only source of knowledge. The problem is that there seems to be a deep-seated dependence on ontological naturalism to the extent that previous demarcation criteria, as well as methods of inquiry, that are supposed to be free from ideological bias are inevitably influenced by it, rendering investigation results ultimately flawed. Ironically, those who fail to see the logical implications of this view assume that inferential justification informs our non-inferential knowledge. In other words, it is assumed that the natural world compels our commitment to ontological naturalism and not the other way around.[2] A shortcoming of this facile rationalization is, perhaps, that it fails to see the real starting point. Consequently, science cannot do without some ontological commitments[3] since our observations and methodologies are only as good as our presuppositions.[4]

    What can we say are our sources of knowledge in interpreting the natural world? Is there enough warrant to believe one frame of reference over another? These questions are loaded with implications, and we do not want mere ideological commitments to be the gatekeepers of the scientific arena. If we truly want to know what nature is made of, and indeed what science seeks to unravel, we need to be careful how we pursue the answers to these fundamental questions.

2   From Criterion to Demarcation

In my estimation, the *Problem of Demarcation* in the philosophy of science is closely related to the *Problem of the Criterion*[5] in epistemology. In developing a suitable criterion of demarcation for science, we first need to identify our sources of knowledge and justified belief. Science is generally thought of as a complete, self-sustaining sys-

---

    [2]The idea that science is the final arbiter in ontology is strongly criticized by philosopher Yvonne Raley. See, for example, Raley (2007).

    [3]Quine's 1948 paper entitled "On What There Is" explains the confusions and difficulties in adopting a particular ontology. Descriptions of qualities, such as an object and its representation in our brains, are either true (realist ontology), or they are not (subjectivist ontology).

    [4]For Popper, the problem of demarcation was that "there cannot be any sharp demarcation between science and metaphysics..." (Popper, 1992b, pg. 161)

    [5]For a more thorough treatment of this topic, see Roderick M. Chisholm's *Theory of Knowledge* (1977), pgs. 6ff.

tem that depends on nothing more than the so-called scientific method of observing, building a hypothesis, making predictions, and testing. It is seldom acknowledged that our tools of observation yield representations that demand subjective interpretation. I am not taking a skeptical position here, but I think it is vitally important to the health of science to recognize when we are putting the cart before the horse, as it were. A criterion of science that does not meet the prerequisite of identifying our sources of knowledge is no criterion at all. For, what does science do without initial statements of fact?

Given our problem of establishing a criterion that could encompass the entire range of scientific disciplines, methodology may be of little use here.[6] How could we apply our criterion to disciplines as diverse as physics and paleontology? The ancient Problem of the Criterion (generally attributed to *Sextus Empiricus*, circa 160–210 AD) stems from our attempt to figure out whether the things we perceive are really as they appear. The problem can be summed up with these two questions:

1. *What* do we know?

2. *How* do we know it?

To understand our perceptions and distinguish true appearances from false appearances, we must employ a criterion (or method) that serves as an aid to distinguish between true appearances and false appearances. However, to develop a criterion we must depend on appearances that we presuppose to be true. The circularity is not difficult to discern. For most things, when we are asked *how* we arrived at certain conclusions, we begin by explaining our inferences as they developed through the experiences that led to them. We seldom think about the presuppositions implanted before our explanations began to take root. This *method-first* strategy for acquiring knowledge is common practice for empirical science, but is it right? Sir Karl Popper's own strategic moves bypassed or ignored the problem, but in doing so, he also excluded genuine science from consideration. I will address this later in the paper.

To resolve the epistemic paradox, one could perhaps identify a *particular* instance of knowledge that requires no method for its justification. In other words, we begin with the first question, *what do we know*, as opposed to the second question, *how do we know it*. When starting with a particular frame of reference, we are not rejecting a criterion for further investigation; we are, in fact, developing it. Do I need additional justification to believe that I am in pain or that I see light? Do I need a criterion to justify such beliefs? Clearly I don't. In both empirical cases, the subject is *prima facie* justified. However, starting with a particular instance of knowledge, or justified belief, does not, in and of itself, constitute science. We have indeed identified a source of knowledge, but science requires a criterion that moves instances of

---

[6]Demarcating science by the "unity of method" remains a mere abstraction that has failed to provide a working criterion.

knowledge to working hypotheses. Moreover, we need to go from instances of knowledge to structuring knowledge into a functional criterion that can work across various disciplines.

Arguably, one of the most successful examples of science, the Scientific Revolution, was one that appreciated and exploited the design characteristics of nature to propose theories that, to some extent, still impact us today. Of course, early philosophers were already writing about the design of our universe as self-evident.[7]

They recognized that there is a natural epistemic dependence on the order and structure of the world, and thus, proposed ideas that cohered with the natural order in the language of mathematics. They acknowledged that it is, in fact, the apprehension of order in nature that determines how nature is understood.

Two competing philosophies on how we are to understand the world in the theory of knowledge, rationalism and empiricism, debate whether knowledge could be justified *a priori* or *a posteriori*. Empiricists base knowledge on sensory experience and induction, while Rationalists base knowledge on reason and deduction.[8] Interestingly, some Empiricists (namely, *logical positivists*) reject a realist ontology and opt for a subjective one devoid of any true picture of reality. Our descriptions, they argue, are mere artifacts of human conventions.[9] Nevertheless, no proposition can function without first presupposing other beliefs about reality, and one cannot continue *ad infinitum* assuming that all belief is inferential.

It is, therefore, logical to propose that all inferential knowledge is subservient to foundational knowledge. This, I believe, is at the core of science. Indeed, when we attempt to answer a question about how we came to a particular conclusion, we want to know the premise on which the conclusion rests. However, any premise that is not a basic premise will suffer under its own need for justification. That is to say, for any subject S to be justified in believing a proposition P on the basis of evidence E, one must be justified in believing E1 on the basis of another proposition E2, and E2 on yet another proposition E3 and so on. If all epistemic justification is inferential, then we wind up with an epistemic regress, or something circular that does not do anything to reinforce our propositions. To illustrate, I may claim to be justified in believing that when I release an apple from my grasp it will not remain suspended in midair, or it will not ascend. I am justified in believing that it will descend because of other known factors, namely, the laws of physics. But, justifying my belief in the laws of physics requires that I know something about the inner workings of physics. That *something* may also depend on something even more fundamental, so that all knowledge is parasitic on how we justify belief. Foundational knowledge, then, serves

---

[7]See, for example, pre-Socratic Greek philosopher, Anaxagoras (ca. 500–428 BC, Apollodorus ap. Diog. Laert. ii. 7); Plato (429–347 BC, Philebus); Stoic philosopher, Epictetus (55–135 AD, Discourses 1.6.1–11); and Saul of Tarsus (a philosopher in his own right c. 5–c. 67 Romans 1:18-20).

[8]Popper turned empiricism on its head by proposing that experience does not verify theories, but rather falsifies them.

[9]See, for example, Rudolf Carnap's *The Logical Syntax of Language* (1937).

as the ground on which to build our propositional pillars.

3   Design as a Criterion of Demarcation

Design as a criterion of demarcation, the proposition expounded in this paper, affirms that design in nature is a properly basic belief[10] and that in order to do science, one cannot escape its constraints. That is to say that science is confined by the boundaries of patterns, order, structure, and regularity that make up the world. Putting it plainly, design bridges the chasm between ontology (*what is*) and epistemology (*how we know it*). This means that every *a posteriori* inference owes its justification to *a priori* knowledge. To be clear, I am not using the term *design* to mean artificiality, plan, or purpose. I am simply using the term to denote order, function, law, regularity, and such characteristics. To use design as a criterion of demarcation of science is simply to let nature's design *characteristics* provide the parameters of investigation.

Accordingly, there is no sense (at least for the advancement of knowledge) in asking whether things in nature have the "appearance" of design when it is almost universally explicitly or tacitly acknowledged in the scientific community.

This design-centric criterion of demarcation affirms design as an ontological feature[11] of the universe, but it does not presuppose causation. Since this view of science is epistemically pre-theoretical, one might say that it is a eutaxiological[12] philosophy of science. For science to flourish, the question of purpose must remain open, though not presuppositionally affirmed. As such, design as a criterion of demarcation is more concerned with *what is* and not necessarily with particular rules for demarcating science. The only rule (and, therefore, our criterion), which involves the search for the degree of order and complexity of processes or structures, [13] is established by the coherence in nature's ontological characteristics. In other words, the activity of scientists is distinguished by the incessant search for comprehensibility, patterns, and things that we recognize immediately without deep ratiocination. *How* these attributes came about is what theories are intended to resolve and, therefore, design is the *sine qua non* of science.

As Popper aptly put it, "[The scientific investigator's] aim is to find *explanatory theories* (if possible, true explanatory theories); that is to say, theories which describe certain structural properties of the world, and which permit us to deduce, with the

---

[10]What I mean by "properly basic belief" is belief that is foundational for knowledge and, therefore, is not dependent on any other epistemic justification. For example, Descartes' own *cogito ergo sum* is a position on what can actually be known from experience, reducing justified belief to the ego that is revealed by the cogito, and this, doubtlessly avoids an infinite regress of justifications. The concept has been around for some time, but has been noticeably popularized by Alvin Plantinga.

[11]Ontological design is the contradistinction to ontological randomness and neutral as to the cause of order.

[12]From the Greek word *eutaxia*, meaning 'good order.'

[13]See Michael Anthony Corey's *God and the New Cosmology: The Anthropic Design Argument* (1993), pg. 10ff.

help of initial conditions, the effects to be explained."[14] This way of reasoning puts teleological explanations on equal footing with teleonomical ones, where both can propose a cause for the effect in question.[15] Design as a criterion of demarcation creates the boundary by which science must operate. It is not to be thought of merely as a theory, but as a determinant from which all theories must operate. The nineteenth century professor of geology and critic of teleological design arguments, Lewis Ezra Hicks, wrote:

> "Physical science is a classified knowledge of external nature; but the possibility of classification, and therefore of science, lies in the fact that there is first a natural, external *order*, whence arises the logical, internal order in the arrangement of facts and principles, which constitutes true science. The external order existed before the science which is based upon it. There was celestial harmony before the science of astronomy was constructed by formulating the laws and principles gathered from observation of the heavens...
>
> This eutaxiological argument, then, seems to have no end to it; for order is universal in nature."[16]

(Hicks, 1883, pg. 17f)

Here we see how this idea of identifying design and searching for the degree of order and complexity of processes or structures can serve to develop a rigorous scientific research program that isn't committed to either teleological or teleonomical presuppositions. As such, the different approaches offered by either side of the aisle are welcome. If our presuppositions force us to commit ourselves to one perspective—one way to approach the same question—we are no longer doing science but engaging in the segregation of thought.

Characteristic of previous demarcation criteria is the failure to provide a direction and a structure from which alternative methodologies can develop and incommensurability can be avoided. Indeed, they have been more restrictive than progressive in their attempts to protect the enterprise from unwelcome company. For instance, Popper's falsifiability criterion only limits scientists on the types of theories they can subject to scrutiny, but it does not imply that other theories aren't true. This is another way of saying that science must proceed methodologically with testable ideas within the margins it has predetermined. But what does it benefit science if all we know is what can be subjected to agreed-upon methods of inquiry? What other

---

[14]See Popper's notes with respect to causal explanations on *The Logic of Scientific Discovery*, pg. 40.

[15]Scientists committed to ontological naturalism have been privileged with monopolizing knowledge without merit and persistently borrow from design to make predictable outcomes.

[16]Hicks's concern was also on the conflation of teleological and eutaxiological design arguments. He did not negate natural order, but the idea that order was indicative of purpose or contrivance.

guidance does falsifiability offer to the scientist? It seems that the filter of science is being misused. If the purpose of the filter is to sieve empiricism from other frames of thought, then it needs to also provide direction as to what the recipe of science is intended to produce.

4 Adoption of an Ontological Commitment

I will not engage in the typical mental exercises of philosophers, or visit their wonderlands and attempt to give life to un-actualized possibilities, Meinong jungles, or other fantasies. Those types of wanderings always seem strange to me as a common sense realist. I am not interested in any form of modal realism, and I do not see its usefulness. Indeed, I have exposed my ontological commitment to *what is* and perhaps by extension, *what is not*. But how do we determine if we are committing ourselves to a proper representation of reality? What factors would help us arrive at our conclusions with confidence? When adopting an ontological commitment, we are faced with the decision of either, as Quine put it, "dulling the edge of Occam's razor," or relying on our crudest observations. It may hold that the limitations of language, or metalanguage,[17] have an impact on our descriptions of reality, but our descriptions continue to produce results because they point to undeniable characteristics of nature. We often use analogies to describe what we are attempting to elucidate, but our symbolic language does not dictate what reality is. The language of science, whatever it may be, helps us to create mental representations of our observations and, thereby, an ontological representation of reality.

In science, however, it is not enough to identify a correct ontology. One must also identify an adequate one because that will ultimately determine our research design. Recognizing the difference between a correct and an adequate ontology will determine how we proceed from our epistemological questions to our methodological ones. For example, a *realist ontology* may be correct, but it would be inadequate as a criterion because it does not give the sort of information that tells us how to proceed with our investigations. My proposition of design as an ontological feature of the universe takes advantage of the characteristics of the natural world as a means to do science. It does not merely define science; it provides a foundation for it. What I am attempting to lay out here is a structure of epistemic justification that would lead to doing good science.

In his analysis of determining the best way for conducting social research, Egon Gotthold Guba (1990), outlines three fundamental questions (Figure 4.1) that help characterize a paradigm. For our purposes, these are questions we should ask when seeking to justify our theories. The first is an *ontological question*: What is the nature of the "knowable"? Or, what is the nature of "reality"? This is the object that we

---

[17]There is a wide range of truth theories; here I am thinking of Alfred Tarski's formulation in which truth statements are determined by their correspondence to reality.

Figure 4.1: Three Fundamental Questions (Guba, 1990)

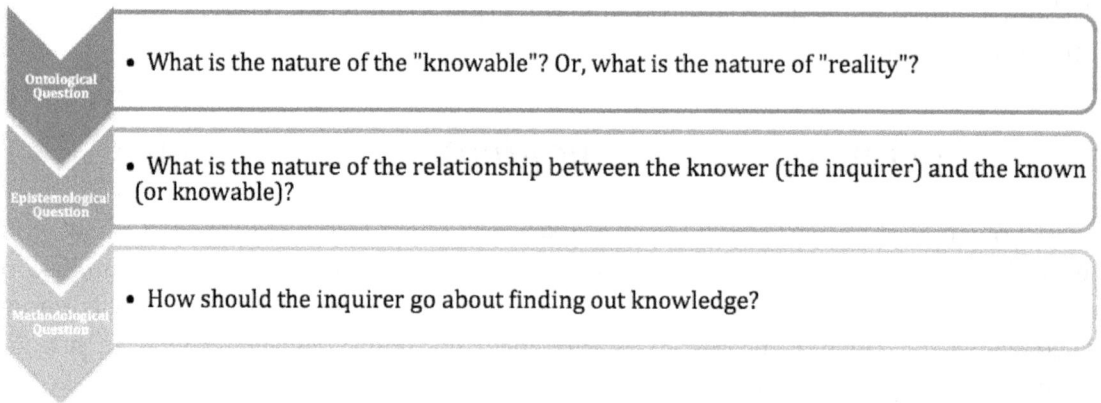

are to subject our methods of inquiry to. If we don't know what we are studying, then we better hope that we obtain knowledge by stumbling upon it in a blind and random search. The second is an *epistemological question*: What is the nature of the relationship between the knower (the inquirer) and the known (or knowable)? Here, our ontological answer guides our epistemological one. Do we espouse objectivism or rationalism? Are our senses the only source of knowledge or can we depend on reason? The third and final question is a *methodological question*: How should the inquirer go about finding out knowledge? Again, the answer to this question lies within the margins of the ontology we accept as true.

5   Fundamental Questions

Design (in the sense that I am referring) is not seen in the works of Austrian-British philosopher, Sir Karl Popper, as far as I am aware of. As a realist, he wanted science to progress in understanding the real world but without ever having the temerity to make any final pronouncements. Popper wrote:

> [T]he system called 'empirical science' is intended to represent only one world: the 'real world' or the **'world of our experience.'** [Emphasis added]
>
> (Popper, 1992a, pg. 14)

But what *is* the "real" world? What is "the world of our experience" according to Popper? What sort of questions would we ask if the "real" world were different? Popper was an anti-conventionalist, so he adopted Alfred Tarski's correspondence theory of knowledge. For clarity, the correspondence theory of *truth*, as it is most commonly known, states that the truth or falsity of a statement is determined only by how it relates to the world and whether it accurately describes or corresponds with it.[18] Popper's criterion was intended to correct our interpretations of what is observed. He believed that statements are not merely fallible, but that they are, in fact, theory-laden. A good example of this is philosopher Paul Draper's position on metaphysical naturalism. For the sake of brevity, I will not cover his view at any length here. Instead, I will introduce you to the idea of metaphysical naturalism as Draper (a prominent advocate) himself defined it during a 2007 interview for the Future of Naturalism conference at the Center for Inquiry in New York. He said,

> Metaphysical naturalism is the view that nature is a closed system. That there are no supernatural entities.

Of course, the most obvious problem for the advocate of methodological naturalism is that he arbitrarily defines what nature is. Once more, it is our presuppositions that drive our methods of inquiry, so whatever our starting point, it should help guide us in our attempts to advance knowledge. If we assume nature is a closed system, as Draper suggested, what kinds of questions are we logically permitted to ask? In our attempt to answer the ontological question, metaphysical naturalism does not have much to say. In other words, it is not very informative to say that nature is natural or physical or that it operates by a nexus of inviolable laws. If we are going to do science, we need to aim our attention on nature's full range of characteristics. In turn, these characteristics should evoke our imagination as to the methods we use in our pursuit of knowledge.

This brings us to another question: How do we trust our cognitive disposition with respect to our perception of the natural world? As alluded to earlier, there are only two ways to answer this question: either we depend on our methods (ignoring their dependence on prior assumptions), or we depend on the reliability of our internal disposition. One could employ an externalist (*reliabilist*) solution and suggest that it is not our *independent* cognitive disposition that we trust, *per se*, but rather how nature actually works. For example, if nature did not have the type of comprehensibility that we could trust, we would be deluding ourselves to think that we can do science. Only order yields predictable outcomes. Our foundational belief, then, is only true if it corresponds with reality.[19] To be sure, it is not possible for a mind to force structure onto an inscrutable or incomprehensible world. Yet, our basic belief is as such because it does not depend on any other knowledge for its justification. It is justified because it is acquired immediately, internally, and objectively. That is to

---

[18]See Robert C. Solomon & Kathleen M. Higgins, *The Big Questions: A Short Introduction to Philosophy 9th Edition* (2013) Glossary pg. 419.

[19]I am not saying that our belief is not justifiable independent of experience, but that it is not true if it does not correspond with it. This is essentially taking account of the so-called Gettier problem as it relates to our mental faculties. My proposition is that our belief is basic and also true and confirmed by our direct acquaintance.

say, we can cogitate upon our direct acquaintance with the structure of the world, which suffices as an objective instance of knowledge or justified belief.[20]

Consequently, the problem we have faced in proposing an adequate ontology, and therefore, a criterion of demarcation between science and non-science is really the problem of determining the sort of explanations we are willing to entertain as we seek to contribute knowledge. Personally, I don't see how we can develop a criterion without presupposing a source for our understanding of things. In order to solve problems, we need to first understand the world. Therefore, if the aim of science is to describe the *real* structure of the world,[21] our immediate reaction—our intuition— tells us that seeking to unravel its design is the way of science. My foundationalist[22] view of science depends on the premise that design is self-evident and, therefore, a properly basic belief. This belief, which is formed referentially by direct acquaintance with the natural order, is a good starting point for science. As I see it, it is perfectly adequate to "demarcate" science by the very thing science is invested in discovering (i.e., its ultimate design).

My position with respect to our perceptions of the natural world differs considerably from that of philosophers, such as Alvin Plantinga,[23] who maintain that teleological design belief is basic. My own position is that only eutaxiological design belief is basic. We may be able to conjoin other basic beliefs about the origin of design, but order, patterns, and the like do not belong to the teleological design classification by default. My opinion is that since ontological questions deal with *what is*, the proper place for design perception is on *attributes*, not *causes*. For instance, if I see a Ford Model T, I may immediately intuit that it is the product of mind, not of chance or necessity. This is perhaps because I am well acquainted with minds and their artifacts and because I, too, possess a mind capable of producing artifacts.

This is true even if secondary causes were employed. However, I am inclined to think that this is not the case with the natural world. I can appreciate order, laws, regularities, and beautiful structures, but I may form a teleonomic belief of nature's design, especially if I am already predisposed to that sort of thinking. I think that an ontological commitment to design attributes is less problematic than presupposing causal connections without drawing those connections by way of inference. Plantinga et al. would have teleological design arguments promoted (or demoted, depending on your attitude toward deductive inferences) to basic beliefs. But I think this is

---

[20]See Roderick M. Chisholm's *Theory of Knowledge* (1977), pg. 7.

[21]Scientific realism is a position that rejects the idea that the world is really a construct of our fertile imagination. The world, according to metaphysical constructivists, is a mere representation of our theorizing. See Peter Godfrey-Smith's *Theory and Reality* (2003), Chapter 12.

[22]I am not going to attempt to thoroughly defend foundationalism from previous criticisms here (i.e., the Agrippa/Münchhausen trilemma, The Gettier problem, etc.) as I think vastly more qualified scholars have adequately responded to them already. See, for example, works by Olaf Tollefsen, Michael DePaul, Richard Fumerton, Laurence Bonjour and Timothy McGrew.

[23]Alvin Plantinga *Where the Conflict Really Lies: Science Religion, & Naturalism* (2011) Chapter 8, pgs. 225–264.

Figure 4.2: Logical Structure of Science

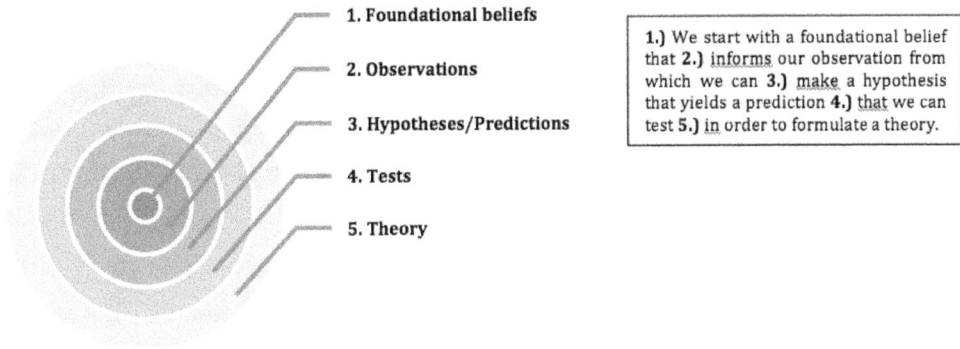

1. Foundational beliefs

2. Observations

3. Hypotheses/Predictions

4. Tests

5. Theory

1.) We start with a foundational belief that **2.)** informs our observation from which we can **3.)** make a hypothesis that yields a prediction **4.)** that we can test **5.)** in order to formulate a theory.

a mistake. Current design arguments have good explicatory power and rightly have a place in science. The problem for design theorists is not so much that they can't make a good case but that they are bringing their case to the wrong court. If the ontological commitments of science are metaphysical naturalism, or physicalism, then design in the teleological sense cannot even step foot in the court.

A criterion is a 'means of judging' and thus, can be used as a standard or a characterizing mark from which we can make judgments. So then, developing a criterion of demarcation concerns questions of how to identify sources of knowledge or justified belief. The structure of epistemic justification proposed here, instead of leading to an infinite regress of explanation, inexorably forces us to respond to the ontological question: *What is?* The picture we get of science, then, does not depend primarily on our observations but, rather, on what sort of ontological commitments predispose our observations to a particular interpretation. The schematic representation in Figure 4.2 shows the logical structure of science as it actually operates.

Popper believed that all knowledge remains fallible and conjectural (Popper, 1992b, pg. xxxv). As such, he did not demand the verifiability of statements; instead, he proposed that statements had to be able to be refuted. He did not believe that we have the capability of giving true descriptions of our observations and, thus, he developed a criterion of demarcating science from non-science by the filtering of falsifiable statements from non-falsifiable ones. In doing so, he thought he avoided an infinite regress of justification and, at the same time, provided a way to keep science moving and knowledge advancing (Popper, 1992a, pg. 26). Since the justification of statements were not judged by their verifiability, but by their falsifiability, falsification required "special rules" in order to refute them. He wrote:

We must clearly distinguish between falsifiability and falsification. We have introduced falsifiability solely as a criterion for the empirical character of a system of statements. As to falsification, special rules must be introduced which will determine under what conditions a system is to be regarded as falsified.

We say that a theory is falsified only if we have accepted basic statements which contradict it (cf. section 11, rule 2). This condition is necessary, but not sufficient; for we have seen that non-reproducible single occurrences are of no significance to science. Thus a few stray basic statements contradicting a theory will hardly induce us to reject it as falsified. *We shall take it as falsified only if we discover a reproducible effect which refutes the theory.* In other words, we only accept the falsification if a low-level empirical hypothesis which describes such an effect is proposed and corroborated. [Emphasis added]

(Popper, 1992a, pg. 66)

It makes one wonder what Popper meant by stating that a theory is taken as falsified only if "a reproducible effect which refutes the theory" is discovered. What sort of effect is reproducible? Is Popper admitting to a necessary condition that must be met before a theory could be refuted? Popper seems to be saying, perhaps inadvertently, that falsifiability is what is minimally needed, and regularity (a design attribute) is what is maximally needed to falsify a theory. Popper avoids the Problem of the Criterion mentioned above as he is not concerned with a particular instance of knowledge; he assumes no knowledge can be gained apart from methodology. By proposing that an empirical scientific system must be refuted by experience, he has given precedence to method over prior instances of knowledge, which are necessary for the development of his criterion in the first place. He goes directly to the second question of the epistemic paradox. We can appreciate the role that experience plays in the justification of statements or claims, but falsifiability imposes an unnecessary burden on science.[24]

It is not difficult to see why falsifiability is not a good criterion of demarcation. Many have been critical of Popper's ideas[25] to varying degrees, but my own concern is that his criterion does not do enough to spur scientific progress. The advancement of science (not just its continuance) does not merely require a demarcation that sets the parameters or scope of investigation; it must also guide as to the sort of characteristics that are to be sought in our pursuits. To this end, my criterion of demarcation stresses that every problem of science should be treated as an engineering endeavor. If design as a criterion of demarcation provides the parameters of investigation, design cannot be *imposed*—as in drawing a target around an arrow—but rather it is to be *discovered*—as in structure, regularities, laws, mechanisms and other like attributes. This immediately excludes the usual suspects such as Marxism, psychologism, astrology, multiverse hypotheses and similar mental notions that stem from fertile imaginations as opposed to careful investigation. Seen this way, design can be used as a way of

---

[24]Hilary Putnam illustrates the burden that falsifiability can put on a theory in his chapter from *The Philosophy of Science* entitled "The Corroboration of Theories," (1991).

[25]Paul Feyerabend, Thomas Kuhn, Imre Lakatos, Max Houck, Larry Laudan, Hilary Putnam, and the Willard Van Orman Quine and Pierre Duhem thesis, to name a few.

disconfirming imposed renderings of reality, just as falsifiability disconfirms theories through the rules of falsification. Moreover, design as a criterion of demarcation does not constitute the notion that our acquaintance with nature informs us of *all* laws, structures, or regularities. It merely gives us the backdrop and, thus, the confidence to operate within nature, encountering anomalies along the way, but informing us just enough to not hinder that confidence. Anomalies are treated as learning from nature its design and our theories are reworked in the exchange of investigation. This may be taken to mean that science is not merely a systematic way to study what nature readily reveals; it is also a way to understand its secrets and limitations.

6   Of Processes and Mechanisms

I submit that conjectures (to use Popper's term) cannot be mere wild speculations, but rather structured inferences aimed at understanding the effect in question.[26] Refutations, as noted earlier, only follow when our expectations are shown to be *imposed* on the natural world, as opposed to *discovered* from the natural world. For example, if we begin with the assumption that nature is a closed system (i.e., metaphysical naturalism), then we will only attribute causal mechanisms to every scientific question. But, clearly this metaphysical presumption assumes too much and results in confusing processes with mechanisms. Although the terms are used interchangeably, confusing terms is always a hindrance to understanding. To be sure, every event results from a process, but not every event results from a mechanism. In other words, a mechanism is always a process, but a process is not always a mechanism. It is perfectly admissible in science to demand processes, but it is not admissible to demand a mechanism, especially when mechanisms are not causally adequate for the effect in question. Mechanism, a term derived from "machine"[27] (a self-contained apparatus or process), limits our options and are wrongly accredited with every phenomenon we encounter.[28] Of course, this does not imply that we should immediately invoke causes that are of the teleological variety, but rather that we need to recognize the limitations foisted upon science as a result of a philosophical bias. A great example that sheds light on the difference between a process and a mechanism is seen in the work of bacterial geneticist James A. Shapiro. His insight on a cell's ability to direct genetic change and repair by means of various complex strategies is one that can be easily missed if we assume that only mechanistic processes are at play. He wrote:

> Another common misperception in many conventional discussions of genomic change is that cells cannot avoid the automatic production of mu-

---

[26] See Hilary Putnam's criticism in Putnam (1991).

[27] From the Greek mēkhanē and the Latin mechanismus.

[28] Distinctions between mechanisms and regularities (i.e. a regularity can be statistical as opposed to deterministic) have also been made by philosophers of science, but these nuances are too vast to cover here. See, for example, Benjamin Barros' "Natural Selection as a Mechanism" (2008).

tations in response to DNA-damaging agents such as UV radiation or mutagenic chemicals. This misperception results from ignorance about the sophisticated apparatus that even the smallest cells possess to repair genome damage and a failure to appreciate the power of cellular genome surveillance and response regimes.

(Shapiro, 1992, pg. 14)

The distinction, as illustrated above, shows how easily we can miss the forest for the trees, as it were, if we assume that all cell change is fatalistically determined as Crick and others have believed.[29] A process such as this requires scientists to look past the assumed mechanism and observe what is happening in real time. In a mechanistic type of scenario, all a scientist needs to do is extrapolate from cause-and-effect assumptions, missing important details as a result. A process that is not mechanistic is lost in history and all that is left is the effect that the process has left behind. Scientists wedded to the idea of a closed system have promoted mechanisms from descriptions of natural phenomena to ultimate causes of all natural phenomena. This attitude, which has exchanged the free enterprise of science for despotism, has limited science in such a way as to create enmity between those that embrace *teleology* and those that embrace *teleonomy*.

In order to eliminate debate or confusion over teleology and teleonomy as it relates to design, we must first recognize the difference between a *cause* and a *process*. While both are empirically discernable, a process is what we observe and a cause is what we infer. In the case of non-mechanistic processes lost in history (i.e., causes that are out of the reach of direct investigation), the only options for investigation would involve reverse engineering or inferences drawn from currently known causal processes. These, of course, are as problematic for teleological explanations as it is for teleonomical ones. Shapiro's insight shows not only how assuming mechanisms were responsible in cell change can yield erroneous results, but also how, absent a mechanism, teleological assumptions (apart from cell cognition) can be equally wrong. Yet, there are instances in which design inferences do yield knowledge that is not derived from assumed mechanisms. Take, for example, the case of so-called "junk DNA." As it turns out, these sequences of DNA that do not code for proteins actually serve other functions (transcription, translational regulation, etc.), but due to a prior commitment to mechanistic processes (as in Crick's Central Dogma), their function had been overlooked. Its function was only later proposed by design proponents committed to teleology.

DNA is an interesting molecule. Since its structure was identified by Watson and Crick in 1953, and the sequence hypothesis proposed by Crick five years later, scientists have been stumped by its sheer elegance and informational properties. There is simply no known mechanism to explain the information embedded in the molecule along its longitudinal axis. The nucleotide base pairs that are sequenced to specify

---

[29]See Francis Crick's "Sequencing Hypothesis" and "Central Dogma" in Crick (1958).

functional roles within the cell are arbitrary, as the sequence does not depend on any affinity between the bases.[30] Again, absent a mechanism, teleological design becomes a very attractive alternative.

In the body of this paper I suggested that my demarcation criterion set teleological explanations on an equal footing with teleonomical ones, where both can propose a cause for the effect in question. I wrote this while being well aware of how I am pitting both law and agency against each other. But, I did this only to make a distinction between two modes of explanation for ontological design, that is, primary and secondary causation, both of which may enjoy the benefits of my demarcation criterion for science. In the first type of explanation, nature may be worked out rationally (*a priori*), and science progresses by way of appealing to cause and effect. By contrast, the second type of explanation (*a posteriori*) deduces knowledge of the natural world from effects to causes.[31] The first may relate laws to mind (top-down), while the second may assume laws to be a mere inherent property of nature (bottom-up). In affirming design as a criterion of demarcation, the distinction set forth here is trivial (not to say superficial). The important thing is that it is laws, affinities, regularities, patterns, etc. that make ontological design self-evident and science possible. Notice here that both terms (teleology and teleonomy) use the prefix *tclco* (from τέλος—télos: end; goal; purpose), and only differ in the suffix, *logy* (from λόγος—logos/logic) and *nomy* (from νόμος— nómos: law). Design is not some abstract idea that requires elaboration; however, I define it here broadly in order to prevent the stalemates that only serve to stifle scientific progress. In proposing ontological design as a criterion, neither teleonomical nor teleological propositions are to be considered as privileged *explanans* and, in this case, immunity is only reserved for the self-evident *explanandum*. What we do not want to do in science is to marginalize ideas that we disagree with simply because they do not conform to how things are usually done.

7 Incommensurability

Science, as we currently know it, is divided not only in focus, in practice, and in language, but also in being able to harmonize natural phenomena across all disciplines. Since there is no single method of science that applies equally to all disciplines, the stratification of science typically depends upon the clear discontinuities that exist in nature (from physics, chemistry, biology, etc.). This expected division makes it

---

[30]Stephen C. Meyer's lengthy book, *Signature in The Cell* (2010), does a great job at shedding light on the problem.

[31]See Karl von Prantl *History of Logic* (1870), concerning German philosopher Albert of Saxony (ca. 1316–1390), who made a distinction between *demonstratio a priori* (the proof from what is before), and *demonstratio a posteriori* (proof from what is after).

difficult to find ways in which all of nature may converge.[32] My proposition is that the one thing that unifies all of science is our dependence on a particular attribute of nature, which is design. It is foundational to every area of science and it is what makes predictions possible. In fact, other criteria of demarcation depend on prior assumptions about the function and structure of the world as well, but the assumptions generally go unnoticed. My proposition of design as a criterion of demarcation also has the benefit of unifying the language of science and resolving incommensurabilities through the sharing of scientific nomenclature with design as its foundation. Language is, more often than not, helpful in converging ideas. Yet in science, where precision is everything, language often becomes an impediment for scientific growth. Since all of science depends on design for understanding and investigating, it also makes sense that my criterion may evoke a unity of scientific parlance not only within specific disciplines, but also across various disciplines with similar objectives.

8   Conclusion

We have veered far from pursuing science for the knowledge it contributes and have only managed to amass academic relationships as we avow to remain loyal to the traditional consensus. The aim of my proposition is to bring together ideas that help us better understand the world. Scientists don't often realize that their preferred ontological commitment drives their scientific methodology and ultimately the sort of results they get. My criterion of demarcation is a demarcation set by the very attributes of nature, so the ontological commitment is one that corresponds with reality. No matter what other philosophical baggage we may bring, here is one undeniable truism:

> The best way to account for the coherence of our experience is to suppose that the outside world corresponds, at least approximately, to the image of it provided by our senses.
>
> (Sokal and Bricmont, 1998)

If design was not a self-evident attribute of nature, science would simply not be possible. We go about our way without a single thought about what keeps our feet firmly planted on the ground as we traverse the plains and vastness of time. Our intuitions inform us with enough acuity that we can go with confidence wherever nature leads, to understand her and lay bare her design. This is science.

---

[32]Physicists seeking a unified "theory of everything" may be on to something, but their focus is typically a reductionistic notion that rests on a mechanistic conception of the world. Perhaps design is a theory of everything. That is, if all of nature exhibits the sort of characteristics that are comprehensible, then it may not be that what we are looking for is the unification of laws, regularities, or order, but rather a meta-principle to rule them all.

References

Barros, B. 2008. Natural selection as a mechanism. *Philosophy of Science* 75(3).

Carnap, R. 1937. *The Logical Syntax of Language*. Open Court Publishing.

Chisholm, R.M. 1977. *Theory of Knowledge*. Prentice Hall.

Corey, M. 1993. *God and the New Cosmology*. Rowman and Littlefield.

Crick, F. 1958. On protein synthesis. *The Symposia of the Society for Experimental Biology* 12:138–163.

Godfrey-Smith, P. 2003. *Theory and Reality: An Introduction to the Philosophy of Science*. University of Chicago Press.

Guba, E.G. 1990. *The Paradigm Dialog*. SAGE Publications.

Hicks, L.E. 1883. *A Critique of Design Arguments*. C. Scribner's Sons, New York.

Meyer, S.C. 2010. *Signature in the Cell: DNA and the Evidence for Intelligent Design*. HarperOne.

Neurath, O. 1959. Protocol sentences. In A.J. Ayer (editor), *Logical Positivism*, pp. 199–208, Free Press, New York.

Plantinga, A. 2011. *Where the Conflict Really Lies: Science Religion, and Naturalism*. Oxford University Press.

Popper, K. 1992a. *The Logic of Scientific Discovery*. Routledge.

Popper, K. 1992b. *Realism and the Aim of Science*. Routledge.

Putnam, H. 1991. The 'corroboration' of theories. In R. Boyd, P. Gasper, and J.D. Trout (editors), *The Philosophy of Science*, pp. 121–138, Bradford Books, 7th edition.

Quine, W.V.O. 1948. On what there is. *Review of Metaphysics* 2.

Raley, Y. 2007. Science and ontology. In *The Proceedings of the Twenty-First World Congress of Philosophy*, volume 12, pp. 143–147.

Shapiro, J. 1992. *Evolution: A View From The 21st Century*. FT Press.

Sokal, A. and Bricmont, J. 1998. *Fashionable Nonsense: Postmodern Intellectuals' Abuse of Science*. Picador.

Solomon, R.C. and Higgins, K.M. 2013. *The Big Questions: A Short Introduction to Philosophy*. Cengage Learning, 9th edition.

von Prantl, K. 1870. *History of Logic*.

Part II

Incorporating Non-Naturalistic Thinking into Academic Study

What are the roles of teleology, theology, and other modes of thought in rigorous academic study? Can non-naturalistic aspects be integrated into mathematical models of causation? What problems can non-naturalistic thinking lead to in academic study? These papers focus on general considerations about integrating non-naturalistic thinking into academics.

5

The Relationship of Bacon, Teleology, and Analogy to the Doctrine of Methodological Naturalism

JAMES C. LEMASTER

Houston Baptist University

Abstract

Francis Bacon divided natural science into physics and metaphysics. He claimed that of Aristotle's four causes, only material and efficient causes belong to the realm of physics, and that final causes, or teleological claims, belong to the realm of metaphysics. Bacon objected to including teleology in physics because in his experience teleological claims tended to discourage the search for efficient causes for natural phenomena. Because Bacon relegated teleology to metaphysics science largely followed his lead, evolving over the next four hundred years a growing distaste for including any teleological implications in scientific explanations. Bacon claimed that human nature, "will yet invent parallels and conjugates and relatives, where no such thing is."

Yet, as the material and efficient causal discoveries by science have progressed since Bacon's time, they have in turn revealed more legitimate parallels and conjugates and relatives than perhaps he could have ever imagined. Stated succinctly, the process of exploring material and efficient causes in nature has also given breathtaking justification for also inferring final causes as well. As such, inferences to teleology in science should be allowed where they are warranted by the empirical evidence.

The tool for determining whether a teleological inference is warranted is analogy. Bacon could have helped science avoid its gradual but inexorable drift into methodological naturalism if he had emphasized how analogy, used as an analytical tool in the process of induction, legitimately leads to reasonable inferences of teleology in nature.

1   Bacon as Pioneer of Methodological Naturalism

Some scholars claim that methodological naturalism traces its early modern roots—at least in part—to Francis Bacon's treatment of final causes. Michael Ruse comments that Bacon "the English philosopher of scientific theory and methodology, ...did not want to deny that God stands behind His design, but Bacon did want to keep this kind of thinking out of his science." (Ruse, 2004, pg. 16) Discussing the genesis of the scientific revolution in the seventeenth century, Harry Lee Poe and Chelsea Mytyk write, "the great surge began with Bacon's proposal for a new disciplined method for the study of the physical world." Bacon's new method, they add, included a clear distinction of two broad causal categories: "Metaphysics speaks to final or ultimate causes, but observations of the world itself tell us about the immediate causes within the world of experience" (Poe and Mytyk, 2007). Cornelius Hunter comments that although Bacon desired scientific practice to seek after explanations of the world that were true, he "also wanted science to restrict itself to naturalistic explanations. Bacon realized that the restriction to naturalism would exclude any realistic, true, explanations that were not strictly naturalistic ...So Bacon ...forfeited completeness." In Bacon's new paradigm, "Science would not investigate all things" (Hunter, 2016).

Indeed, Bacon stated, quite clearly, his goal to exclude final causes—or teleological explanations—from the realm of physics and instead to assign them to metaphysics:

> Physic handles that which is most inherent in matter and therefore transitory, and Metaphysic that which is more abstracted and fixed. And again, that Physic supposes in nature only a being and moving and natural necessity; whereas Metaphysic supposes also a mind and idea ...The true difference between them must be drawn from the nature of the causes that they inquire into ...Physic inquires and handles the Material and Efficient Causes, Metaphysic the Formal and Final

> (Bacon, 1905, pg. 459).

What were Bacon's reasons for limiting teleological inferences or explanations to the realm of metaphysics and excluding them from the more practical realm of experimental, empirical, and inductive physics?

2   Bacon's Relegation of Teleology to Metaphysics

As Francis Bacon looked back on history, he recognized and lamented that for many centuries, in studying nature, people have often uncritically followed a tendency to appeal to final causes. This tendency, Bacon insisted, "is so far from being beneficial, that it even corrupts the sciences" (Bacon, 1902, 2:2, pg. 51).

2.1 The Idol of the Theater

One of the factors upon which he placed much of the blame for this corruption was the long-lasting effects of the most famous ancient Greek approaches to natural philosophy. Repudiating an ancient tendency to erect what he labeled 'the idol of the theater,' Bacon censured several ancient Greek thinkers for neglecting diligent, thorough, empirical experiment and observation, and for too quickly and too frequently letting sophistry and superstition guide their interpretation of nature. "Aristotle ...had already decided", Bacon remonstrated, "without having properly consulted experience as the basis of his decisions and axioms, and after having so decided, he drags experiment along as a captive constrained to accommodate herself to his decisions" (Bacon, 1902, 1:48, pg. 18). Bacon then broadened his criticism: "The disputatious and sophistic school entraps the understanding, while the fanciful, bombastic, and, as it were, poetical school, rather flatters it. There is a clear example of this among the Greeks, especially in Pythagoras, ...but it is more dangerous and refined in Plato and his school" (Bacon, 1902, 1:65, pg. 19). Bacon complained, "This evil is found also in some branches of other systems of philosophy, where it introduces abstracted forms, final and first causes, omitting frequently the intermediate [causes] and the like" (Bacon, 1902, 1:65, pg. 19).

Bacon especially repudiated what he saw as a tendency in the past (whether by ancient Greek natural philosophers or by their more recent Christian counterparts) to jump far too quickly, based upon scanty experimental data, to broad universal principles and, from then on, to judge nature deductively by those philosophically-driven generalizations.

> We allow that the ancients had a particular form of investigation and discovery, and their writings show it. But it was of such a nature, that they immediately flew from a few instances and particulars ...to the most general conclusions or the principles of the sciences, and then by their intermediate propositions deduced their inferior conclusions, and tried them by the test of the immovable and settled truth of the first...Their flying off to generalities ruined everything

(Bacon, 1902, 1:125, pg. 47).

Bacon also chided those in more recent times (such as alchemists) who had followed the misguided precedent of those influential Greeks. They too practiced "the premature and forward haste of the understanding, and its jumping or flying to generalities and the principles of things" (Bacon, 1902, 1:64, pgs. 18–19). Thus, Bacon not only blamed ancient Greek thinkers for the sorry state of natural philosophy, but he also unflinchingly criticizes some natural philosophers of the post-Greek Christian era:

> The corruption of philosophy by the mixing of it up with superstition and theology, is of a much wider extent, and is most injurious to it ...Some of

the moderns have indulged this folly with such consummate inconsiderateness, that they have endeavored to build a system of natural philosophy on the first chapter of Genesis, the book of Job, and other parts of Scripture

(Bacon, 1902, 1:65, pg. 19).

Within this criticism of Christian natural philosophers from the past and until his day, Bacon seems to imply that they have succumbed too easily to the influence of their ancient, renowned Greek predecessors. Thus, perhaps for him, the most fundamental blame still lies with the Greeks.

2.2   The Idol of the Tribe

Bacon also places blame for the backward state of natural philosophy that had persisted for so long on another idol—what he calls "the idol of the tribe." This category subsumed what he saw as the many serious prejudices and flaws in perception and reasoning with which humans seem to be naturally and universally plagued. Besides flowing from the influence of ancient Greek thinkers, Bacon asserted that the hasty, overly-deductive method he decried "has always been done to the present time from the natural bent of the understanding" (Bacon, 1902, 1:104, pg. 38). More specifically, Bacon said, "The idols of the tribe ...arise either from the uniformity of the constitution of man's spirit, or its prejudices, or its limited faculties or restless agitation, or from the interference of the passions, or the incompetence of the senses, or the mode of their impressions" (Bacon, 1902, 1:52, pg. 15).

Clearly, Bacon saw many serious defects in mankind's inherent perceptive reasoning abilities. There is one, however, that deserves special attention for the purposes of this article. That defect, according to Bacon, is the tendency to be tricked into perceiving far more purposeful similarities, patterns, and order in nature than are actually there. He contended that "the human understanding, from its peculiar nature, easily supposes a greater degree of order and equality in things than it really finds " (Bacon, 1902, 1:45, pg. 13).

It is also possible that Bacon felt that humans tend to erroneously identify mechanistic operations in nature. He wrote, "The human understanding is perverted by observing the power of mechanical arts, in which bodies are very materially changed by composition or separation, and is induced to suppose that something similar takes place in the universal nature of things" (Bacon, 1902, 1:66, pg. 19). He may be charging that since we humans are so familiar with mechanical devices—and the teleology that underlies their manufacture— that we mistakenly attribute similar mechanical teleology to complex phenomena, organisms, or structures we observe in nature. In other words, we insist that an analogy exists between natural phenomena and humanly-manufactured artifacts when such an analogy is not justified.[1]

---

[1]David Hume later voiced a similar dismissal of analogies between nature and artifacts. This dismissal will be addressed later in the paper.

Scattered in Bacon's writings there lies another possible clue as to why Bacon had such a pessimistic view of the reliability of human understanding, including how humans understand nature. He seemed to have accepted as an unquestionable given, a vast, perhaps even categorical, gulf between human thinking methods, patterns, and capacities and those of God. He asserted that some defenders of "the Christian religion ...celebrate the union of faith and the senses as though it were legitimate ...and gratify men's pleasing minds ...but in the meantime *confound most improperly things divine and human*" (emphasis mine) (Bacon, 1902, 1:89, pg. 33). Bacon deprecated such efforts, claiming that "not only fantastical philosophy, but heretical religion spring from the absurd mixture of matters divine and human. It is therefore most wise soberly to render unto faith the things that are faith's" (Bacon, 1902, 1:65, pg. 19).

Bacon also mentions two other "idols" that have kept mankind from accurately comprehending the world: that of the "den" and that of the "market." However, it seems that for Bacon, it is primarily through the idols of the theater and of the tribe that the human tendency to seek final causes has caused so much trouble (Kennington, 2004, pg. 6). In what specific ways, then, did Bacon believe that inferring teleology in nature spoils the scientific process? Why did Bacon assign final causes to the realm of metaphysics? Why was he loath to include them, along with material and efficient causes, in the experimental practice of physics (as he called it)?

Two important negative consequences that Bacon claims have resulted from teleological inferences in natural philosophy deserve special attention. First, including such inferences retards the progress of investigation, evaluation, and discovery in natural philosophy:

> Final causes [are] misplaced ...This misplacing hath caused a deficience, or at least a great improficience in the sciences themselves. For the handling of final causes, mixed with the rest in physical inquiries, hath intercepted the severe and diligent inquiry of all real and physical causes, and given men the occasion to stay upon these satisfactory and specious causes, to the great arrest and prejudice of further discovery

(Bacon, 1863, pg. 94).

Elsewhere, Bacon charges that final causes are nothing "but remoras [obstructions] and hindrances to stay and slug the ship from further sailing; and have brought this to pass, that the search of the physical causes hath been neglected and passed in silence" (Bacon, 1863, pg. 94). These reasons closely parallel those given by current philosopher of science Robert Pennock as he defends methodological naturalism: "Once such supernatural explanation are permitted ...all empirical investigation could cease, for scientists would have a ready-made answer for everything" (Pennock, 1999, pg. 292).

A second harmful consequence that Bacon feared is that mankind's overactive imagination in inferring final causes will lead us to a false picture of reality. As I mentioned above, Bacon was convinced that humans erroneously see far more purposeful order, resemblances, and relationships in nature than actually exist: "The human understanding ...will yet invent parallels and conjugates and relatives, where no such thing is" (Bacon, 1902, 1:45, pg. 13). Bacon went a bit further, claiming that teleological thinking in natural philosophy projects purely human traits upon nature, and as a result, grossly misrepresents reality:

> Although the greatest generalities in nature must be positive, just as they are found, and in fact not causable, yet the human understanding, incapable of resting, seeks for something more intelligible, ...namely, final causes; [however] ...they are clearly more allied to man's own nature, than the system of the universe, and from this source they have wonderfully corrupted philosophy.
>
> (Bacon, 1902, 1:48, pg. 14)

In summary, it seems that Bacon's reasoning process was as follows: Man is prone to anthropocentrically inject his own kind of teleology into nature. Human and the Divinity possess altogether distinct modes and capacities of thinking. Therefore, Man must exercise extreme caution to avoid misreading nature according to his own habits and understanding. As a result, one very practical way of exercising this caution is to exclude the invocation of final causes from explanations in the physical sciences.

As he laid out his broad proposals for a new and improved scientific method, was Bacon justified in omitting final causes from experimental science or did he misstep? Can hindsight help expose flaws in his approach that have produced adverse consequences for our correct understanding of the natural world? Can scientific experience gained through the intervening years since Bacon point out healthy corrections to his methodological omission of teleology from physics?

2.3   Responding to Bacon's Views on Teleology

Some current philosophers and scientists may give credit to (and perhaps applaud) Francis Bacon for planting the roots of methodological naturalism through his exclusion of final causes from physics. However, Bacon's treatment of teleology in science has a number of shortcomings, and those today who utilize reasoning similar to Bacon's may be mistaken as well.

First, one needs to remember that one of Bacon's reasons for removing final causes from physics was his severe antipathy to what he viewed as Greek quasi-pantheistic notions of final causes existing within physical things via nature alone. In one sense Bacon seems to have been railing against the eminent Greek philosophers, not because of their invocation of teleology *per se*, but because of the particular version of teleology he saw them propagating. Bacon says in *De Augmentis*:

[Aristotle and Plato] were perpetually inculcating them [final causes]. Though in this respect, Aristotle is more culpable than Plato; as dropping God, the fountain of final causes, and substituting nature in his stead, receiving final causes through his affection to logic, not theology ...Aristotle, had no need of a God, after having once impregnated nature with final causes, and laid it down that nature does nothing in vain; always obtains her ends, when obstacles are removed.

(Bacon, 1815, pgs. 112–113)

In this passage, Bacon says what bothered him about Greek teleology was that it seemed to him that Aristotle ignored a supernatural source of final causes (the God of Bacon's Christianity) and by substitution, credited nature's components and laws themselves—or Nature itself in a comprehensive sense—with a semi-pantheistic intentionality.

How Bacon responded to Aristotle's version of final causes was where his first misstep regarding teleology may have occurred. Bacon was disenchanted with both the method and the results of Aristotle's version of natural philosophy, laden as Bacon saw it with final causes grounded in nature independent of God. Therefore, he perhaps too hastily or generally reasoned that all teleological inferences from nature—Greek or not—were illegitimately imposed by biased philosophical presuppositions. Therefore for him, teleology could not justifiably be inferred within the due course of his kind of gradual, data-driven, experimental, empirical, inductive methodology. Bacon himself is probably guilty of throwing out too much of the teleological baby with the ancient philosophical bathwater.

Bacon's second problem is that he expressed views on teleology, which upon thoughtful consideration, are difficult to reconcile with each other. Despite his rejection of final causes in physics, Bacon continued to staunchly affirm the validity and utility of teleological inferences, either within metaphysics or at a more abstract level. For him, God was somehow guiding nature with His own final causes.

One troublesome problem with Bacon's writings, is that even despite holding a generally Christian worldview, he did not clarify well his own convictions of how God's governing teleology was apparent in the physical world in any detectible ways. For example, regarding the juxtaposition of physical causes and final causes, Bacon cautions,

Keeping their precincts and borders, men are extremely deceived if they think there is an enmity or repugnancy at all between them ...Both causes [are] ...true and compatible, the one declaring an intention, the other a consequence only. Neither doth this call in question or derogate from Divine Providence, but highly confirm and exalt it.

(Bacon, 1863, pg. 95)

Bacon also asserted,

> Any one who properly considers the subject will find natural philosophy
> to be, after the Word of God, the surest remedy against superstition, and
> the most approved support of faith. She is, therefore, rightly bestowed
> upon religion as a most faithful attendant, for the one exhibits the will
> and the other the power of God.

(Bacon, 1902, 1:89, pg. 33)

The perplexing question is, how can Bacon support his definitive assertions that
natural philosophy "declares," "confirms," "exalts," or "exhibits" anything about a
divine Being if teleology is not detectible (or at least not allowed to be inferred) within
the realm of the physical sciences?

In the 1920s, philosopher C.D. Broad summarized Bacon's views on teleology
as follows:

> Bacon holds that the existence of teleology in Nature is an obvious fact,
> and that the investigation of final causes is a perfectly legitimate branch
> of Natural Philosophy. It has, however, been misplaced; for it belongs to
> the division of Natural Philosophy which Bacon calls Metaphysics and not
> to that which he calls Physics ...[Bacon believed] that there is no art of
> Applied Teleology as there is an art of Applied Physics. Now Bacon holds
> that the existence and some of the attributes of God can be established
> conclusively by reflexion on the teleology of Nature.

(Broad, 1926)

Again, however, the bothersome question is, how can the existence and any
attributes of God be established conclusively by reflection on nature's teleology, when
teleology has been excluded from categories of explanatory causes available in the
physical sciences? Either the physical nature of the universe is actually telling us
something about teleology or it isn't. It is hard to escape the feeling that Bacon, or
his interpreters, are trying to defend both options at the same time. In summary,
Bacon's persistent claims that natural philosophy somehow confirms and exhibits the
final causes of a Creator seem to lack support once he relegates teleology to the realm
of metaphysics and excludes it from physics.

Bacon's third problem is that he did not supply evidence to support his repeated claims that teleological thinking in natural philosophy falsely projects purely
human traits upon nature. Despite the human propensity for bias, why couldn't
the physical sciences themselves yield legitimate evidence sufficient to persuade us
that what appear to be compelling indicators of final causation are actual and not
merely illusory? Could Bacon have been too pessimistic about the shortcomings of
the human tendency to notice patterns, order, and analogies in nature?

Closely related to Bacon's unsupported perspective about human understanding is another unsupported perspective about God that was raised above. Bacon does not make obvious in his writings the specific reasons he was certain that God's thinking methods, patterns, and capacities belong in a vastly distinct realm than mankind's. For him, there just seemed to be a virtually unbridgeable gap between those two kinds of methods, patterns, and capacities, and Bacon does not appear to have explained in detail why he held such a view. Kennington comments that for Bacon, "the final cause is an anthropomorphism ...The natural desire of the human understanding is not for knowledge, but for assurance or reassurance that all that is is for the sake of a purpose or for the good. In the sequel it is assimilated to religion or the desire for a benevolent and comprehensive divine power" (Kennington, 2004, pg. 21). If Kennington is right, then Bacon thought that this anthropomorphism is an illusion (or delusion?). This, therefore, seems to imply that it is also an illusion that the benevolent and comprehensive divine power governing the purported teleology is highly analogous to us humans. If Bacon believed this—and one thesis of this paper is that he did—he does not seem to have explained why he believed it. If it is merely an assumption of his, it is an assumption that makes all the difference for our topic of methodological naturalism and the legitimacy of teleological inferences in science.

There is a fourth problem with Bacon's views. It was shown above that Bacon reproached the tendencies in natural philosophy that preceded him to hastily jump to broad generalizations from a paucity of experimental evidence. Instead, he recommended a highly inductive scientific method that "constructs its axioms from the senses and particulars, by ascending continually and gradually, till it finally arrives at the most general axioms" (Bacon, 1902, 1:19, pg. 9). Bacon nowhere gives evidence as to why the method of diligent, continual, gradual, thorough ascent from empirical observations of nature (such as he advocates) will not, or even in principle could not, cause scientists to arrive at general explanations (or axioms) that include or infer teleology.

A fifth problem (and one for which he could not be blamed in the 1600s) is that Bacon's assertions that teleological inferences are presumptuous, obstructive, misplaced, or specious within the physical sciences seem increasingly out-of-step with scientific developments since his time and especially with recent major trends. In part, it appears that Bacon disparaged long-standing natural philosophy, because during the centuries since the time of the ancient Greeks, much more empirical knowledge had accumulated, overturning many traditional notions about how the world really operated. With just a hint of superiority Bacon remarks, "In that age [i.e., the age of the famous ancient Greek philosophers] the knowledge both of time and of the world was confined and meagre, which is one of the worst evils for those who rely entirely on experience ...In our times ...the mass of experiments has been infinitely increased" (Bacon, 1902, 1:72, pg 23).

Approximately five hundred years have passed since Bacon penned his works urging a new scientific method. In that time span the mass of experiments, and the knowledge of the natural world has increased at least as much as what Bacon appealed to here. Five centuries of scientific insights have accumulated through an inductive (experience-rich) process that would likely meet his approval. As the next section will show, those insights have yielded evidence—to borrow Kennington's phraseology—of intermediate causes in the ascent that render the presence of final causes highly plausible (Kennington, 2004, pg. 21), if not compelling.

Many, if not all of the problems just listed could be resolved if Bacon had possessed two more things: First, a fuller grasp of the power and legitimacy of analogy as a reasoning tool, and second, five hundred more years of greatly broadened and deepened empirical understanding of biological organisms along with the corresponding remarkable advancements in humanly-designed technological devices and systems. Many of these advances in both science and technology persuasively indicate that through the prudent use of analogy between the two—for example analogy from organs to human devices and vice versa—teleology of natural phenomena can validly be inductively inferred. Moreover, those same phenomena, importantly including biological organs, organisms, systems, and their interactions can also be much more accurately and holistically (and thus better) understood. Before looking at analogy in more depth, it is worth briefly addressing the topics of intrinsic and extrinsic teleology, and their relevance to analogical inferences about nature, which might have encouraged Bacon not to exclude them from the process of reasoning within physics.

3   The Relationship Between Intrinsic and Extrinsic Teleology and Analogy

Certain scholars and I would agree that Bacon took a serious misstep when he excluded final causes (i.e., teleology) from science. We all would agree that it would behoove science to reinstate what seems obvious to many close observers of nature: that teleology is a real component of many natural systems or organisms. Yet, some of these scholars would undoubtedly take issue with my assertion that analogy from organs to artifacts serves as a legitimate tool for better understanding the biological world. For example, Edward Feser and Alexander Sich assert a rigid distinction between intrinsic and extrinsic teleology. They assert that teleology within natural objects or organisms is intrinsic and oppose using an artefactual, extrinsic model to explain it. They reject attempts to pattern our understanding of natural systems and organisms (or organs) after humanly-designed artifacts[2] (Feser, 2011; Sich, 2012).

Upon further reflection, however, the strict intrinsic-extrinsic distinction expounded by Feser and Sich can become significantly less clear than advertised. While on one hand, claiming that causes in natural (and unconscious) components (e.g.,

---

[2]Sich asserts this distinction, in part, because he sees a deep flaw in invoking machine analogies for understanding biological organelles. Sich claims such reasoning emerges from an illegitimate reversal of "Aristotle's crucially important principle: Art(ifact) imitates nature".

those within biological organisms) have "built-in," or "inherent"(Feser, 2011), teleology, Feser also insists that all teleology requires an "intellect" to direct it (Feser, 2008, pgs. 115–116). Since unconsciousness and a directing intellect seem to be incompatible in the same entity, it follows that in most cases, teleology in nature ultimately requires an extrinsic intellect, even if temporarily, or "mediately," such teleology is intrinsic.

Thomas Aquinas clearly illustrates this mix of ultimate extrinsic and mediate intrinsic teleology in his famous Fifth Way by using the analogy of an arrow shot toward the target by the archer (Aquinas, 1920). Once the arrow leaves the bow, the direct teleology of the archer transfers into the arrow itself. The "principle of motion and change" has morphed from extrinsic (originating in the archer) to intrinsic (the arrow itself). So, are final causes throughout most of nature intrinsic or extrinsic? The answer seems to depend upon which stage in the overall process one focuses on.

An intrinsic-extrinsic distinction also gets muddy when one appeals to Darwinian processes to explain biological diversity. Evolutionary biologists often invoke "co-option" as the evolutionary process that caused the development of highly complex organs or systems. Yet in Darwinian co-option, "traits that had evolved under one set of conditions were co-opted [via mutation and natural selection, the latter being extrinsic to the organism] to serve a different function under a second set of conditions" (McLennan, 2008). Thus, how could one strictly claim that the new function—being at least partially generated from the outside—is "built-in," "intrinsic," or "immanent"? Indeed, could not the entire evolutionary process—in sum total, a purported plethora of functional changes over time—be viewed, at least in part, as a vast collection of extrinsic teleological transformations?

Intelligent design theorist William Dembski also writes that the intrinsic-extrinsic distinction, exemplified in Aristotle's distinction between natural objects and artifacts "is prone to a certain fuzziness." Dembski suggests that recent scientific findings might point to "external design" (i.e., to an ultimately extrinsic source of what we only presently see as intrinsic teleology, what Dembski seems to closely associate with information) displayed in living things (Dembski, 2014, pg. 55).

If the boundary between intrinsic and extrinsic teleology turns out to be a distinction without a crucial difference, then the analogical gap between artifacts and natural objects seems to narrow. Moreover, the use of an analogy between biological organelles and humanly-built devices or systems seems even more legitimate for more fully and accurately understanding the former.

3.1   Artifacts vs. Natural Objects

Through the use of analogies between natural objects and human artifacts, could Bacon have legitimately kept teleology within the realm of physics, and is such an approach possible and even advisable within science today? As alluded to above, scholars like Feser and Sich would roundly reject such a methodology. For them, there is a marked distinction between natural objects and artifacts, and the latter should never be invoked—presumably even analogically—as a means to describe the former.

Sich asserts, "Biology studies natural living things," and then clarifies that the term "natural" applies to those things whose final causes (or "principles for motion/change") are intrinsic. Conversely, artifacts are those things with extrinsic final causes and thus lie outside the purview of biology (Sich, 2012). Sich highlights "Aristotle's crucially important principle: Art(ifact) imitates nature, and not the other way around." He then warns that numerous scientists and thinkers, including intelligent design theorists, claim rather that "nature imitates art(ifact)", which is a "'mechanistic' reductionist error" (Sich, 2012, pg. 53). Sich concludes that references to living cells, or their components, as "factories" and "machines" are examples of "muddled thinking that flips reality on its head" (Sich, 2012, pg. 54).

Whether any Intelligent Design proponent has ever claimed that nature imitates artifact is open to debate.[3] Intelligent design advocates' viewpoints often seem to imply merely that certain natural structures or systems and certain artifacts seem analogous. Analogy leaves open the question of whether either analogue derives from the other. Therefore, it does not seem that intelligent design-style reasoning in any way threatens to reverse Aristotle's common sense dictum.

Another point favoring the legitimacy of drawing analogies between natural objects and human devices or systems concerns the range of manifestations of teleology. Feser (apparently following Aristotle and Aquinas) defines final causes so broadly that they can range from the merely regularly observed, unconscious, and superficially nonfunctional effects[4] to the highly specified, complex, pre-contrived, and goal-directed strategies of the human mind. Presumably, because of this broad range, Feser seems to allow for a kind of continuum according to which teleology in nature (the physical world) is more-or-less obvious, more-or-less clear. Feser (2008, pg. 194) says, "The human mind manifests final causality more obviously than anything else ...The mind is the clearest paradigm of final causality"[5].

If humans, and especially our minds, are the clearest manifestation of teleology in the natural world, it would seem thoroughly legitimate to use human teleology—

---

[3]William Dembski (2014, pg. 60) claims that "intelligent design aims to discover solid scientific evidence of real teleology in nature. Typically, it applies some formulation of the materialist-refuting logic to supply such evidence. But that does not mean intelligent design is committed to a mechanistic or reductionistic or artefactual view of life or the universe".

[4]Linked to a point made previously, Feser (2008, pg. 238) asserts that Aristotelians even view unconscious entities like asteroids or mountains as exhibiting "final causality insofar as [they are] ...'directed toward' the production of some determinate range of effects."

[5]Feser (2008, pg. 248) also implies a graduated continuum exists when he writes, "Human thought and action are the most obvious examples of phenomena that exhibit irreducible teleology, but they are far from the only ones. Indeed, final causality pervades the natural world *from the level of* complex biological organs *all the way down to* the simplest causal interactions at the microscopic level" (emphasis mine).

especially as manifested in the things it produces—to analogically analyze teleology in nature. We want to continue to grow in our understanding of the natural world. We already understand the human world quite well. As I will mention below, analogy is a key learning tool that helps us use what we already know well to begin to understand what we do not yet know. Therefore, there is much to commend the use of teleologically-rich human artefactual devices and systems to make (and potentially test) analogical inferences about what appear to be highly similar, teleologically-rich features in the biological world.

All this being said, Stephen Talbott raises some formidable empirical challenges to the legitimacy of such artifact-organism analogies. Talbott correctly points out that the scientific literature abounds with references to cellular organelles as "machines" (Talbott, 2010b, pg. 28). In contrast, Talbott extensively illustrates that the picture emerging from cellular biology is not one of a fully bottom-up, mechanistic realm. Talbott asserts that the explanation of life as "inanimate, molecular-level machinery was misconceived" (Talbott, 2010a, pg. 9). Rather, a better understanding is one of "the unified and irreducible functioning of the cell and organism as a whole—a living, metamorphosing form ..." (Talbott, 2010a, pg. 9) Talbott summarizes in the broadest terms how to understand organisms: "Everything ...in the cell, is ...a manifestation of life, ...As a metaphor for the scientific understanding of biology, ...'it's life all the way down'." (Talbott, 2010a, pg. 14)

Talbott's descriptions of the newest findings demand significant rethinking of how best to understand and explain life at the cellular level. Yet, they do not necessarily rule out the appropriateness or usefulness of making analogical connections between human devices and systems and organic life at the cellular level.

First, the attempts by Talbott to generally summarize what we are learning about the fundamental nature of organisms do not seem particularly helpful. They seem perilously close to the tautology, "Life is living." At least using analogies from human design allows us to learn something new. One of the powers of analogy lies in its delicate balance of similarity and distinction. In order for an analogy to work best, the two analogues must be highly similar, but not identical. Stating that "A raven is black like a raven" doesn't clarify much.

Second, referencing F. R. Lillie, Talbott asserts, "it is hardly possible for an unchanging complex to explain an ordered developmental stream. Constant things cannot by themselves explain dynamic processes" (Talbott, 2010a, pg. 8). However, it is possible for constant things to explain parts of a dynamic process. Analogies with humanly-made devices should not be made to help us understand the whole system of organic life, but they should serve quite well to help us understand certain components of life.

Finally, in a section where Talbott seems to want to summarize the central claim that "The Organism is not a Machine," he says, "there is no obvious similarity between a sewing machine or clock or any other machine and, say, a twisting, gesturing chromosome" (Talbott, 2010b, pgs. 36–37). For some reason, Talbott has chosen overly simplistic and largely irrelevant examples that are bound to make analogies with human devices sound implausible. There are examples from the world of human invention and design that are more on the cutting-edge and at least somewhat more similar to the dynamic, interactive phenomena for which Talbott is seeking an explanation.

The latter part of this article presents some of these much more highly relevant examples. Before proceeding with those examples, it is important to provide an introductory overview of analogy. The next section proceeds to answer the following questions: What are some principles governing analogical reasoning, and what are some characteristics of useful, highly informative analogies?

4   Analogy (or, Inferring Teleology Inductively)

4.1   Analogy Theory: A Brief Introduction

Broadly speaking, analogy is one of the most fundamental ways we humans learn new things. It is a method that reasons from the scope of what we already know toward a better understanding of things in the larger scope of what we do not yet know. More specifically, analogy is a tool that has been widely used and proven useful both in science, broadly, as well as within hypotheses concerning the origins of biological novelty (LeMaster, 2014).[6] However, to begin, what standards cultivate the proper use of analogy in scientific hypotheses to help foster scientific advancement and steer scientific investigation closer to the truth? What are some guidelines to encourage the use of plausible and strong analogies and avoid frivolous and weak ones?

Dedre Gentner and Arthur Markman emphasize that the sharing of abstract relationships (which they describe as "higher order") between two analogues is more important than other more concrete (or mundane) similarities between them: "A matching set of relations interconnected by higher order constraining relations makes a better analogical match than an equal number of matching relations that are unconnected to each other" (Gentner and Markman, 1997, pg. 47).

A teleological analogy seems to meet this standard of mapped higher-order relations. A causal relation is required between teleology and specified and complex features in human artifacts. That same higher-order causal relation is then mapped onto the specified and complex features in natural phenomena, suggesting teleology as an analogous causal factor.

A large quantity of overlapping features also may not matter to the strength of the analogy: "A theory based on the mere relative numbers of shared and non-shared predicates cannot provide an adequate account of analogy, nor, therefore, a sufficient basis for a general account of relatedness" (Gentner, 1983, pg. 156). C. Kenneth Wa-

---

[6]Except where direct or indirect quotes from others are documented, most of the Analogy section of this paper is copied or adapted from my dissertation.

ters criticizes "enumerative inductions" portrayed as good analogies. Waters laments that these kinds of portrayals

> ignore the very relations that make the arguments analogical. Hence, traditional accounts fail to capture the special pattern of reasoning underlying analogical inferences. No wonder these traditional accounts have prompted philosophers to conclude that analogical arguments are too weak to justify scientific hypotheses and to belittle the justificatory role played by them throughout the history of science.[7].
>
> (Waters, 1986, pg. 503)

One could ask, "Doesn't invoking a teleological cause require a literal similarity—not merely an *analogical* one (by Gentner's terminology)—between human artifacts and biological organisms?" Proponents of teleology in nature (e.g., intelligent design theorists) are not necessarily claiming literal similarity in the physical manifestations of the two things being compared. However, in the most important aspect, specified complexity (a higher order aspect), artifacts, and organisms are literally similar, even isomorphic. Moreover, physical similarity is not necessary when comparing certain human artifacts—say, a tractor and a copy of *Hamlet*—in order to justify inferring that both exhibit teleology. One may point out all the ways a tractor and *Hamlet* lack "literal similarity" if one wishes, but they both include the same general, higher order cause, namely teleology.

At a general level, Holyoak and Thagard claim that "analogy is part of inference to coherent explanatory theories" (Holyoak and Thagard, 1995, pg. 170). While they ultimately prefer Darwin's theory (Holyoak and Thagard, 1995, pg. 174), they still praise William Paley's famous watchmaker analogy for teleology in nature, calling it "a sophisticated analogy based on a system mapping" (Holyoak and Thagard, 1995, pg. 172). They continue:

> The argument from design clearly uses a system mapping, as shown by the ...higher-order 'cause' relation in both the source analog (watch/watchmaker) and the target analog (world/God). So there is indeed an analogical explanation that makes the hypothesis of divine creation a legitimate candidate to be evaluated by inference to the best explanation.
>
> (Holyoak and Thagard, 1995, pg. 174)

All this being said, Holyoak and Thagard claim that analogies, even those containing apparent isomorphisms, are not guaranteed to perfectly reflect reality. They conclude only that given isomorphism, analogies drawn from the source "will have some plausibility for the target." Since plausibility falls significantly short of certainty,

---

[7]Weitzenfeld labeled as "partially defective" analogies based only on "number of properties in common or ...degree of similarity" (Weitzenfeld, 1984, pg. 137)

they claim the best one can do is use the analogy tentatively in order to "to generate inferences about the target, and then check whether these inferences actually hold up when the target domain is directly investigated" (Holyoak and Thagard, 1995, pg. 30). Such a process sounds like one of which Bacon would probably approve.

4.2 Problems with Hume's Criticism of Analogical Design Arguments

In *Dialogues Concerning Natural Religion* David Hume discusses analogies used to infer teleology in nature (Hume, 1948, pg. 17). Hume's criticism of such analogical reasoning is well-known: "wherever you depart, in the least, from the similarity of the cases, you diminish proportionably the evidence; and may at last bring it to a very weak analogy, which is confessedly liable to error and uncertainty" (Hume, 1948, pg. 51). Hume then suggests that due to striking dissimilarities, the analogies between human artifacts and natural phenomena may end up being no better than "a guess, a conjecture, a presumption concerning a similar cause" (Hume, 1948, pg. 51).

Although one could raise several objections to Hume's reasoning, a particular objection deserves special attention. When Hume points out that as similarities lessen, analogies weaken, he neglects to mention that the converse is also true: as the two analogues become more and more similar, the analogy grows stronger. Over two hundred years of both technological developments and scientific discoveries have emerged since Hume's day. Many of those developments and discoveries have prompted (and arguably are increasingly prompting) ever-more-plausible analogical inferences between human artifacts and natural structures or systems than those Hume criticized in *Dialogues*. For example, the similarities between the specified complex information in DNA and that within computer languages[8] or the similarities between cellular machinery and man-made micro-machinery probably seem to suggest analogically parallel causation more powerfully than anything Hume (or Bacon) could have imagined.

4.3 Science Regularly and Successfully Uses Analogies

Scientists frequently use analogies as powerful tools in the process of developing and analyzing the plausibility of their hypotheses. Einstein and Infeld commented,

It has often happened in physics that an essential advance was achieved

---

[8]Hubert Yockey (1981, pg. 16) writes, "It is important to understand that we are not reasoning by analogy. The sequence hypothesis applies directly to the protein and the genetic text as well as to written language and therefore the treatment is mathematically identical." Contrary to Yockey's claim, when considered materially, the two systems he compares (functional amino acid chains and human languages) may indeed be analogous, not strictly identical. However, the core feature of specified complex information maps identically between the two, justifying an isomorphic teleological analogy for the causes of both.

by carrying out a consistent analogy between apparently unrelated phenomena ...To discover some essential common features, hidden beneath a surface of external differences, to form, on this basis, a new successful theory, is important creative work.

(Einstei and Infeld, 1938, pgs. 286–287)

Johannes Kepler purportedly called analogies his "most trustworthy masters" (Polya, 1954, pg. 12). Gentner lists examples of scientists such as Newton, Galileo and Rutherford using analogies in the development of their theories (Gentner, 1982; Hume, 1948; Chamberlin, 1965). Henri Poincaré felt that analogy was crucial to guide scientific progress (Poincaré, 1907). Raimo Anttila, linguistics professor at UCLA, claims, "Analogies are utterly essential parts of all theories ...[Analogy] is particularly valuable when the object of investigation is not directly observable" (Anttilla, 1977, pg. 17). Holyoak and Thagard provide a long list of analogies which proved eminently fruitful in the progress of science (Holyoak and Thagard, 1995, pgs. 186–187). Others have noted analogy's crucial role in forming and verifying scientific hypotheses (Woltering, 2012; Gust, Krummack, Kühnberger, and Schwering, 2008).

Holyoak and Thagard do caution that analogy is "often apt to lead to false conclusions" (Holyoak and Thagard, 1995, pg. 190). They therefore suggest that analogy should only be used for the discovery and development phases of the typical scientific process, and not resorted to when doing evaluation, although they note that some well-respected scientists have not met their criterion. They specifically call attention to Darwin, who "was explicit in listing natural selection/artificial selection analogy as one of the grounds for belief in his theory" (Holyoak and Thagard, 1995, pg. 190).

4.4   Compelling Current Examples of Teleological Analogy

Advances in a field called synthetic biology is one place to find increasing analogues for undergirding an inductive teleological argument. Molecular biologists and synthetic biologists have already made impressive strides toward producing microscopic life in the laboratory. Two characteristics of these synthetic biology experiments deserve special attention. First, the features of the products they produce increasingly resemble those of living bacteria. Second, the component of intentional, intelligent design thoroughly permeates these experiments and the products they produce. What synthetic life pioneer Craig Venter and his company have been doing for the last few decades is a paragon of a teleological phenomenon.[9]

Chemist Fazale Rana outlines three broad stages—all permeated with intelligence and intention of teleology—in the Venter team's project for synthesizing life:

---

[9]This venture has at least two final ends: 1) to create life in the lab that can potentially benefit mankind; 2) to make lots of money.

a) preparing for the genome synthesis, b) synthesizing the genome, and c) transplanting the genome. Interestingly, a fourth stage—again intensively directed by teleology—was also likely required for Venter and his team to produce their most recent breakthrough (New Scientist, 2016). That fourth step would be minimizing the genome.

Each stage in this project critically requires purposeful aspects such as intensive planning, close attention to detail, thorough collaboration, critical feedback and consultation, evaluation and learning from mistakes, creativity, immense reasoning power, and experience, all carried out with dedication to a clear end goal. In short, from beginning to end, teleology is an indispensable component of the process.

In addition to the direct involvement of Venter's team, their success, to date, has also required highly controlled laboratory conditions, as well as specialized equipment, which itself was specially prepared by other teleologically-guided human corporations. Moreover, the team has been building upon the generations of previous researchers whose successes required similar kinds of teleologically-based resources.[10] Fazale Rana summarizes that Venter and his colleagues

> depended on the accomplishments of the scientists who came before them. The technology to chemically synthesize oligonucleotides represents a remarkable technological accomplishment resulting from the dedicated efforts over the last half-century of some of the best scientists in the world (including Nobel laureates). Without these brilliant minds and remarkable achievements, Venter's team would have had no hope to carry out the total synthesis of the *M. genitalium* genome.

(Rana, 2011)

In summary, an intelligent and intentional process has produced a functioning genome, isomorphically analogous to the countless other genomes we have observed in the natural world. The product or effect is essentially identical to the two analogues. We know that the causes of one of those effects (the synthetic genome) requires intensively teleological efforts and processes from beginning to end. Sound analogical reasoning, beginning from empirically observed phenomena (the teleologically-driven efforts of the synthetic life team), inductively provides a reasonable explanation of how the analogous effects (the vast multitude of natural genomes we observe) came to be.

Another example of an organelle that undergirds an inductive teleological argument is the ATP synthase. This organelle exhibits a highly isomorphic analogy to a hydroelectric turbine generator. For example, regarding integrated functionality in both analogues, rotary motion, two motors, a central drive shaft and a stator, along with several subunits (Nakamoto, Scanlon, and Al-Shawi, 2008; Block, 1997; Seelert,

---

[10]For example, Murray and Szostak (1983) comment, "The availability of recombinant DNA technology allows... the construction of artificial chromosomes from cloned fragments of DNA."

Poetsch, Dencher, Engel, Stahlberg, and Mülller, 2000), act in concert to convert potential energy to mechanical energy and then to a final, transportable energy form (a generator produces electrical energy, an ATP synthase produces ATP).

Given its features, many molecular biologists have candidly likened the ATP synthase to humanly-designed devices. Ballmoos says, "The F1F0 ATP synthase is a miniature engine composed of two opposing rotary motors" (von Ballmoos, Wiedenmann, and Dimroth, 2009, pg. 655). Block says, "when the crystal structure of the F1-ATPase was eventually solved, it looked every bit like a three-piston rotary motor, with a hexagonal ring of $\alpha - \beta$ pairs surrounding a drive shaft made up of the γ-subunit" (Block, 1997). Interestingly, Nakamoto et al. (2008) highlights how analogy to human motors may have played a key role in understanding the ATP synthase: "The central location the c subunit and its obvious resemblance to a camshaft stimulated investigators to develop approaches that would demonstrate rotation."[11]

In addition, while lacking physical resemblance, there are some recent humanly-developed devices or systems (or collections thereof) that exhibit (or likely will soon exhibit) at least some of the features that cause Steve Talbott to reject the organelle/machine analogy. Such features may include interactions of both top-down and bottom-up causation, sensitivity and adjustment to the immediate surrounding physical context, context-driven regulation on the fly, and remote, purposeful interactions. Examples of devices or systems exhibiting at least some of these features are the Cassini spacecraft (which has undergone two extensions to its original mission), airplane autopilot systems, smart bombs, smart phones, Apple's Siri, GPS navigation systems and cars that parallel park themselves, which hopefully culminates in safe self-driving cars complete with "radar-based cruise control, motion sensors, lane-change warning devices, electronic stability control and satellite-based digital mapping" (Jet Propulsion Laboratory, 2016; The Automotive eZine, 2016).

Arguably, cases like these would supply Bacon with the tools he would need to present empirically based evidence that ascends to an explanation of an important natural phenomenon while simultaneously supporting his claims that natural philosophy confirms and exalts the truth of final causes (according to Bacon, those of a divine Providence) (Bacon, 1863, pg. 95).

5   Conclusion

From our perspective in the twenty-first century, it seems that Francis Bacon painted himself into a corner. On one hand, one can sympathize with his conclusion that Aristotelian intrinsic final causes may have strongly contributed to long-standing scientific stagnation. One can also affirm his affection for a system that strongly emphasizes induction through gradual and thorough observations and experiments of causes and

---

[11]Interestingly, Piccolino (2000) includes side-by-side illustrations of a water wheel and the ATP synthase, in Figure 3, in his article about molecular machines.

effects. However, it seems that Bacon went too far by excluding all teleology outright from the realm of physics. This then leaves him in a conundrum because without recourse to teleology that we can detect from our empirical observations of nature, his Christian claims that natural philosophy still exalts, confirms, and exhibits what for him was the source of teleology (God), sounds like nothing but unsupported dogma that must be accepted purely on the word of some human religious authority.

If Bacon were alive today, one might advise him to take another course: Retain your empirical, inductive method, Francis, but don't necessarily forbid teleological explanations in the physical sciences if inductive inferences from that storehouse of empirical observations merit such explanations. The reason is that since the early 1600s, science has steadily accumulated evidence (empirical, observed data from many branches of science) from which we often can inductively present explanations that include teleology and that seem to best explain what we observe. How would we do it, you ask? Practically how would we carry out such a hybrid of your empirical inductive method along with the teleological causes when merited? By analogy, Francis. Use analogy as induction's friend.

References

Anttilla, R. 1977. *Analogy*, volume 10 of *Trends in Linguistics: State-of-the-Art Reports*. The Hague: Mouton Publishers.

Aquinas, T. 1920. *Summa Theologica, Question 2. "The existence of God", Article 3. "Whether God exists?"*.
http://www.newadvent.org/summa/1002.htm#article1

Bacon, F. 1815. *De Augmentis Scientiarum Volume I*. M. Jones Paternoster-Row, London.

Bacon, F. 1863. *The Two Books of the Proficience and Advancement of Learning*. Parker, Son, and Bourn, London, fourth edition edition.

Bacon, F. 1902. *Novum Organum*. P.F. Collier, New York.

Bacon, F. 1905. De dignitate et augmentis scientiarum. In J.M. Robertson (editor), *The Philosophical Works of Francis Bacon*, George Routledge and Sons, London.

Block, S.M. 1997. Real engines of creation. *Nature* 386(6622):217–219.

Broad, C.D. 1926. *The Philosophy of Francis Bacon: An Address Delivered at Cambridge on the occasion of the Bacon Tercentenary 5 October 1926*. Cambridge University Press, Cambridge, MA.
http://www.ditext.com/broad/bacon.html

Chamberlin, T.C. 1965. The method of multiple working hypotheses. *Science New Series* 148(3671):754–759.

Dembski, W.A. 2014. *Being as Communion: A Metaphysics of Information*. Ashgate Publishing, Burlington, VT.

Einstei, A. and Infeld, L. 1938. *The Evolution of Physics: From Early Concept to Relativity and Quanta*. Simon and Schuster, New York.

Feser, E. 2008. *The Last Superstition: A Refutation of the New Atheism*. St. Augustine's Press, South Bend, IN.

Feser, E. 2011. Nature versus art. *Edward Feser Blog* .
http://edwardfeser.blogspot.com/2011/04/nature-versus-art.html

Gentner, D. 1982. Are scientific analogies metaphors? In D.S. Miall (editor), *Metaphors: Problems and Perspectives*, pp. 106–132, Humanities Press, Atlantic Highlands, NJ.

Gentner, D. 1983. Structure mapping: A theoretical framework for analogy. *Cognitive Science* 7:155–170.

Gentner, D. and Markman, A.B. 1997. Structure mapping in analogy and similarity. *American Psychologist* 52(1):45–56.

Gust, H., Krummack, U., Kühnberger, K.U., and Schwering, A. 2008. Analogical reasoning: A core of cognition. *Künstliche Intelligenz* 22(1):8–12.

Holyoak, K.J. and Thagard, P. 1995. *Mental Leaps: Analogy in Creative Thought.* MIT Press, Cambridge, MA.

Hume, D. 1948. *Dialogues Concerning Natural Religion.* Hafner, New York.

Hunter, C. 2016. What evolutionists don't understand about methodological naturalism. *Darwin's God Blog* .
http://darwins-god.blogspot.com/2011/07/what-evolutionists-dont-understand.html

Jet Propulsion Laboratory. 2016. About the mission. *Cassini Mission to Saturn* .
https://saturn.jpl.nasa.gov/mission/about-the-mission/overview/

Kennington, R. 2004. *On Modern Origins: Essays in Early Modern Philosophy.* Lexington Books, Lanham, MD.

LeMaster, J. 2014. *A Critique of the Rejection of Intelligent Design as a Scientific Hypothesis by Elliott Sober from his Book Evidence and Evolution.* Southern Baptist Theological Seminary, PhD dissertation.

McLennan, D.A. 2008. The concept of co-option: Why evolution often looks miraculous. *Evolution, Education, and Outreach* 1(3):247–258.

Murray, A.W. and Szostak, J.W. 1983. Construction of artificial chromosomes in yeast. *Nature* 305(5931):189–193.

Nakamoto, R.K., Scanlon, J.A.B., and Al-Shawi, M.K. 2008. The rotary mechanism of the atp synthase. *Archives of Biochemistry and Biophysics* 476(1):43–50.

New Scientist. 2016. Artificial cell designed in lab1 reveals genes essential to life .

Pennock, R. 1999. *Tower of Babel: The Evidence against the New Creationism.* MIT Press, Cambridge, MA.

Piccolino, M. 2000. Biological machines: From mills to molecules. *Nature Reviews Molecular Cell Biology* 1(2).

Poe, H.L. and Mytyk, C.R. 2007. From scientific method to methodological naturalism: The evolution of an idea. *Perspectives on Science and Christian Faith* 59(3):213–218.

Poincaré, H. 1907. *The Value of Science*. The Science Press, New York.

Polya, G. 1954. *Induction and Analogy in Mathematics*, volume 1 of *Mathematics and Plausible Reasoning*. Princeton University Press, Princeton, NJ.

Rana, F. 2011. *Creating Life in the Lab*. Grand Rapids: Baker Books.

Ruse, M. 2004. The argument from design: A brief history. In W.A. Dembski and M. Ruse (editors), *Debating Design: From Darwin to DNA*, pp. 13–31, Cambridge University Press, Cambridge.

Seelert, H., Poetsch, A., Dencher, N.A., Engel, A., Stahlberg, H., and Mülller, D.J. 2000. Proton-powered turbine of a plant motor. *Nature* 405(6785):418–419.

Sich, A. 2012. The independence and proper roles of engineering and metaphysics in support of an integrated understanding of god's creation. In *Engineering and the Ultimate: An Interdisciplinary Investigation of Order and Design in Nature and Craft*, pp. 39–59, Blyth Institute Press, Broken Arrow, OK.

Talbott, S. 2010a. Getting over the code delusion. *The New Atlantis* pp. 27–51.

Talbott, S. 2010b. The unbearable wholeness of beings. *The New Atlantis* pp. 3–27.

The Automotive eZine. 2016. The gps self-driving car .
http://automotive.lilithezine.com/Self-Driving-Car.html

von Ballmoos, C., Wiedenmann, A., and Dimroth, P. 2009. Essentials for atp synthesis by f1f0 atp synthases. *The Annual Review of Biochemistry* 78:649–672.

Waters, C.K. 1986. Taking analogical inference seriously: Darwin's argument from artificial selection. In *Contributed Papers*, volume 1 of *PSA: Proceedings of the Biennial Meeting of the Philosophy of Science Association*, pp. 502–513.

Weitzenfeld, J.S. 1984. Valid reasoning by analogy. *Philosophy of Science* 51(1):137–149.

Woltering, J.M. 2012. From lizard to snake; behind the evolution of an extreme body plan. *Current Genomics* 13(4):289–299.

Yockey, H.P. 1981. Self organization origin of life scenarios and information theory. *Journal of Theoretical Biology* 91(1):13–31.

6 ‖ Applied Theology: Exploring The Utility of Theological Method in Scientific Research with Genomic Research as an Example

JAMES D. JOHANSEN

Liberty University

1 Introduction

Are there things that cannot be explained adequately by science? How can one know the capability of one's measuring tools without an external calibration source and an absolute frame of reference to provide a basis for what one is considering? Looking at these kinds of questions, this paper introduces the claim that theological method can be useful in science. It discusses the research being done to address these kinds of questions, summarizes the general approach being used, and points out preliminary observations.

It is difficult to understand the challenging trends in the various areas of scientific development and continue to discover greater insight at the forefront within a number of these fields. The claim being explored in this research suggests that adding theology into aspects of scientific research can help. There are examples of scientists who have done this but perhaps not in a structured way. This does not mean that utilizing the scientific method is not useful. Instead this should be considered as one of several tools that can be used in research and discovery.

A key premise is there are several ways non-naturalistic methodology improves science: 1) It can mitigate science's philosophical limitations and shed light on what

Figure 6.1: Intersection of Theology and Science Research

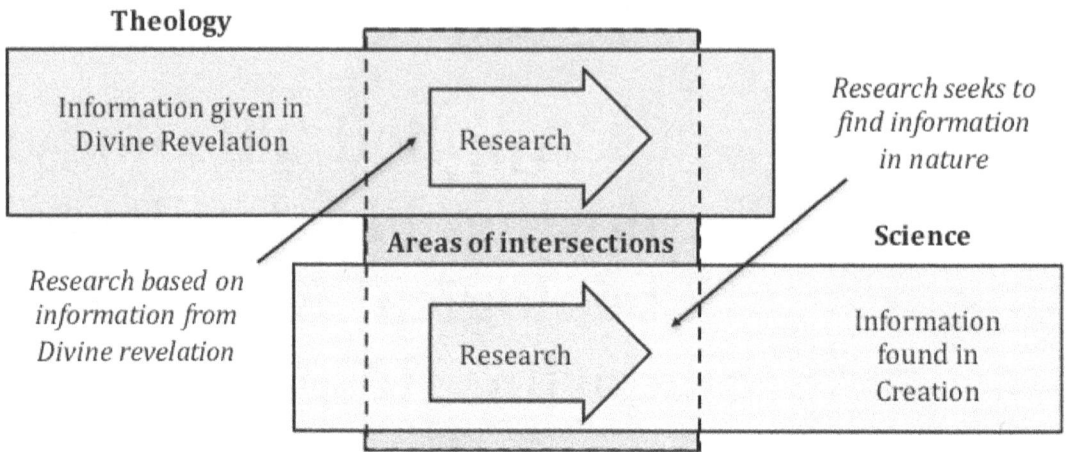

some would call its blind spot. 2) It can expand the possible solution space domain and provide an approach to help down select viable explanatory alternatives. 3) It can provide more tool diversity and allow for greater checks and balances in formulating experimental designs, analyzing data, and interpreting the results.

How can theology be directly applied? Theology and science do not easily mix in many ways since their primary foci serve very different things. The study of God and how the study of God can draw man closer to God is different from the study of the physical world, and the discovery of new natural phenomena happens without an owner's manual to know how nature works. Still, there are intersections between the two disciplines, and, at boundary conditions, one may need to switch to a larger scope allowed by a theologically-enhanced domain that includes both physical and nonphysical phenomena for better explanatory perspectives.

Theology begins with Divine revelation (i.e., information) and then does research, while science does research and experimentation in order to uncover information about the physical world. There is an intersection where these two could work together in the research domain as illustrated in Figure 6.1. Perhaps there is a way of linking Divine revelation as a basis for setting up research and using this as one of the tools to help discover the nature of the information found in the natural world.

By utilizing the benefits of both domains, this paper shows that greater insight can be made into reality and the nature of the scientific world. Part of this will utilize insight from philosophy of science and philosophy of theology. Philosophy can act as a mirror while theology can act as a measuring stick.

By utilizing both domains one can ask whether what is being suggested aligns with the realm of Divine truth with both physical and non-physical reality. Theology provides a basis for truth and knowledge claims, and it provides a sanity check. If we allow theology to be brought into the discussion, then we can consider areas that go beyond what we can fully know. There are some guidelines provided from theology that can be delineated by utilizing theological method, and these principles can be applied as augmentation to scientific analysis. If all that is displayed around us is not completely knowable, then allowing divine revelation to give us a basis for understanding is useful. It is also claimed that with theology one can more precisely define the characteristics of a designed world

If God does exist and all is not revealed or knowable by man, theology can highlight the differences between what we know and what is true reality. Thus, theological principles can help provide a basis for determining truth and what is knowable. By effectively having a yard stick to measure the world around us, it may highlight science's shortcomings and suggest where scientific research may be profitably done.

Several theologians and how their ideas relate will be considered in this chapter. For example, Clark points out specific areas where he argues theology can help science, and Poythress uses both his theological and mathematical training to consider evidences of divine characteristics found in the created world. This is an interesting concept that is worth considering further and is utilized in this work.

Since theological and scientific methods are proposed to work together, it is useful to briefly introduce both for the context of this paper. First, theological method is similar in philosophical concept to the scientific method. Theological method defines the approach one takes to study theology and make determinations on the nature of God, the ultimate source of truth, Divine authority, and an assessment of biblical content. Defining a structured theological process can enable an individual to explain to others how their theological analysis can be a rigorous process that can identify objective truth. The realm of truth can extend beyond empirical data from measured results. A general theological method is not as common as the ubiquitously used scientific method. Divine revelation is given and there is a direct examination of this information. There are different views on the authority of Divine revelation.

Second, the scientific method is a concept that most would be aware of by name. Scientific research looks at empirical data without making use of an owner's manual to understand how it is meant to work. This is challenging and limits what one can measure under the circumstances that the measurements were taken. Genomic research is a good example of an active area of research where new associations are being made between functions and specific aspects of organisms' genomes. Scientific examples in this chapter include 1) bioinformatics and genome sequence assessment, 2) DNA multi functionality and polymorphisms, and 3) what may be called "system biology," looking at cellular function as a system.

2   Claim: Theological Method in Scientific Discovery Has Utility

The claim of this chapter, and the research that is supporting it, is that augmenting scientific research with theological method can improve scientific reasoning. Looking at a specific scientific area, utilization of theology can improve science in the examination of genomic information. It may help 1) generate better approaches for experiment design, 2) allow for more robust data interpretation, and 3) improve the calibration of what can be known.

In support of this claim one can consider what some call the limitation of existing methodological paradigm. The current naturalistic arguments do not adequately explain the evidence seen in the world around us. Even as science advances there are signs that we cannot converge on a complete understanding with empirical tools and data only. From a microbiology standpoint John Sanford argues, "Isn't it remarkable that the Primary Axiom of biological evolution essentially claims that typographical errors and limited selective copying within an instruction manual can transform a wagon into a spaceship in the absence of any intelligence, purpose, or design? Do you find this concept credible?" (Sanford, 2015, pgs. 2–4)

What does theological method add to the argument and what impact might it make? David Clark suggests five areas of impact:

1. Christian theology gives an explanation of why the universe is orderly, why it is prone to mathematical interpretation, why it exists, and why its existence makes a difference.

2. Christian theology provides a metaphysical foundation for the rational justification of science. Theology explains why natural science is achievable as a rational endeavor.

3. Christian theology explains why science makes a difference. Christian thinking provides a reasoning for our conclusions that knowledge gained through science is valuable.

4. A growing subgroup argues theology can speak to science by directing future research. This goes past offering metaphysical, rational, and axiological grounds for science.

5. Christians might use what they know from theology to assist in assessing one scientific theory over another. Consider a situation where scientists evaluate two incompatible models, both of which appeal to and account for substantial portions of the same empirical data. Theology can help in this area (Clark, 2010, pgs. 254–256).

3   Claim Justification: Literature Review Where Theological Method May Help

The approach for claim justification used here is twofold. First, I will examine examples of scientific research where the arguments presented in the literature may provide a gateway for using theological insights to provide a more compelling explanation of their findings. Second, I will look at examples of scientific research that allow for both scientific and theological methods to work together profitably in their research.

The literature shows several categories of scientists and theologians that include a theological component to their work:

1. Scientists who are open-minded and willing to honestly consider non-natural sources of truth, like Anthony Flew towards the end of his life.

2. Scientists who are Christians and have a desire to tie their faith into science like John Sanford.

3. Scientists who allow for Divine intervention, choosing to work with the scientific method but allowing for nonphysical truth and inputs, like some researchers with ENCODE and intelligent design proponents like Stephen Meyer, Ann Gauger, Doug Axe, William Dembski, and Robert Marks.

4. Theologians who explore science drawing from theological truth that gives more insight into the physical world, such as Vern Poythress, David Clark, and Carl Henry.

Figure 6.2 illustrates what could happen as a result of this cooperative engagement. As developed by Paul Nelson, this is an example of the use of triangulation in scientific exploration (Nelson, 2016). If theology and science can help each other, then perhaps there is a triangulation concept that may be applicable. The scientific method, using methodological naturalism, is in one corner, and theology is in another corner. Moving up the scientific path is intelligent design. This assumes a minimal integration of theological principles. More specific theological integration moves science up further on the science scale.

This section suggests that the chapter's claim can be applied to the information in the genome. The argument is that its origin cannot be adequately explained by naturalistic evolutionary processes. Naturalistic evolutionary processes only allow for a limited tool set to explain results, and researchers try to uncover the meaning in the data under these limitations. Because of these limitations, the scientific method alone does not always produce the best explanatory interpretation or the most compelling result. Plus, as is the case for scientific discovery, theories can be proven wrong. For example, there is a growing body of evidence that many things have been incorrectly described as junk DNA. Instead, significant portions of the genome that

Figure 6.2: Science and Theology Integration

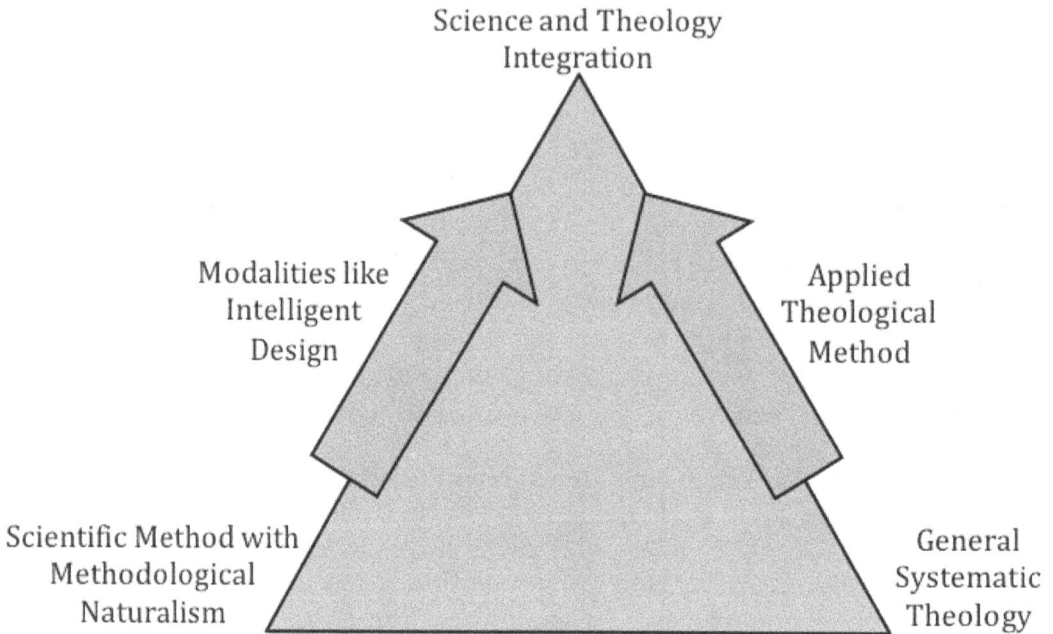

were considered junk have been found to perform a useful function. Now, mapping function of DNA previously described as junk is an active area of research.

3.1 Experiments Based Upon Natural Selection Tenets

Two examples of research are discussed below illustrating scientific results that do not easily align with the tenets of natural selection and common descent. Thus there is the possibility of applying the theological method to expand the scope of explanations of the results.

The Encyclopedia of DNA Elements (ENCODE) Project Consortium in 2007 published its identification and analysis of functional elements in 1% of the human genome utilizing the ENCODE pilot project. The results do not align with natural selection. According to an early publication of its findings in *Nature* there were some interesting and unexpected developments. In general, the goals of the project were to provide biologically-informed examples of the operation of the human genome by using high-throughput methods to identify and catalogue the functional elements encoded. In their pilot stage, 35 groups offered over 200 experimental and computational data sets that examined 29,998 kilobases (kb) of the human genome. This set of 30 Mb(about 1% of the human genome) was large and diverse enough for beta testing of multiple experimental and computational methods. The 30 Mb set was from 44 genomic regions, with about 15 Mb contained in 14 regions where considerable biological knowledge exists, and the rest was from 30 regions selected via a random-sampling method (The ENCODE Project Consortium, 2007).

In the article they mentioned the following:

> [We] uncovered some surprises that challenge the current dogma on biological mechanisms. The generation of numerous intercalated transcripts spanning the majority of the genome has been repeatedly suggested, but this phenomenon has been met with mixed opinions about the biological importance of these transcripts. Our analyses of numerous orthogonal data sets firmly establish the presence of these transcripts, and thus the simple view of the genome as having a defined set of isolated loci transcribed independently does not seem to be accurate.

(The ENCODE Project Consortium, 2007)

Additionally in the article, they stated that "...we have also encountered a remarkable excess of experimentally identified functional elements lacking evolutionary constraint, and these cannot be dismissed for technical reasons"(The ENCODE Project Consortium, 2007). Thus the ENCODE project found results that challenged the existing paradigms and was not aligning with the existing natural selection dogma.

A second example looks at the lack of genomic findings that align with the common descent concept of the Last Universal Common Ancestor (LUCA). From an experimental point of view, one would want to find evidence of common decent all the way down to the genomic level. How would this look? There should be common threads in DNA, but the results do not align with such a premise. One would expect to find common functional coding at genomic level. There should be common functional DNA coding across species for things like flight. Yet, there are multiple mechanisms for flight and no common genetic code for flight. In regards to common decent, if all life emerged from a common core one should be able to find a very simple common core that would align with LUCA. Unfortunately, discoveries show a very complex core that does not suggest a common simple progenitor organism following Darwinian methods (Forterre and Philippe, 1999).

What would happen if an objective scientist looked at the data and he formulated a theory that embraced all of the evidence no matter where that might lead? Anthony Flew could be a good example of this. He was a staunch atheist for most of his life. Yet, what he could not avoid, since he was an open-minded scientist, was where the implications of astronomy and cosmology lead him. As a result, he came to the following conclusion when he was confronted by the totality of the facts:

> Although I was once sharply critical of the argument to design, I have since come to see that, when correctly formulated, this argument constitutes a persuasive case for the existence of God. Developments in two areas in particular have led me to this conclusion. The first is the question of the

origin of the laws of nature and the related insights of eminent modern scientists. The second is the question of the origin of life and reproduction.

(Flew, 2008, pg. 95)

Using only the scientific method, Flew found there were things being conveyed that cannot be quantified by the physical world. Thus there is a need to inject what I would argue is the theological method. So Flew argued the following:

> Science qua science cannot furnish an argument for God's existence. But the three items of evidence we have considered in this volume—the laws of nature, life with its teleological organization, and the existence of the universe—can only be explained in the light of an Intelligence that explains both its own existence and that of the world. Such a discovery of the Divine does not come through experiments and equations, but through an understanding of the structures they unveil and map.
>
> (Flew, 2008, pg. 155)

3.2   Researchers Embracing a Christian Worldview

One can look at a number of Christian scholars who are willing to look deeper at the implications of science when they have a dual science and theology focus. Two examples used here are John Sanford, a Christian biologist, and Vern Poythress, a scholar in both theology and mathematics. Then, utilizing Poythress' arguments, this section will discuss the implications of augmenting science with theology to answer scientific issues.

John Sanford has had a long career that involves seeing scientific evidence as supporting more than naturalistic results. He suggests that it is unlikely that we will ever completely understand the complexity of the genome, and that we most likely cannot completely understand it.

> The bottom line is this: the genome's set of instructions is not a simple, static, linear array of letters—but is dynamic, self-regulating, and multi-dimensional. There is no human information system that can even begin to compare to it. The genome's highest levels of complexity and inter-action are probably beyond the reach of our understanding, yet we can at least acknowledge that these higher levels of information exist. While the linear information within the human genome is extremely impressive, the non-linear information must obviously be much greater. Given the unsurpassed complexity of life, this has to be true.
>
> (Sanford, 2015, pgs. 9–11)

Vern Poythress has a PhD in Theology and a PhD in mathematics. As a result of his expertise in both these fields, he has the ability to consider the intersection of theological and mathematical truth and how these can come together in creation. He, therefore, sees evidence of God's divine characteristics manifested in creation. He notes that three characteristics of natural law that correspond to attributes of God: 1) God is omnipresent, present in all locations, 2) God is immutable, He does not change, and 3) God is eternal, present at all times. It is not an accident that these three attributes of God are mirrored in scientific law (Poythress, 2014).

This forms a useful paradigm where the realm of theology and its study of God can find aspects of correspondence in the realm of science and its study of nature. If God created the world around us, it is obvious there would be a reflection of His character in what He has made. The fact that elements of His handiwork that reflect His nature are detectable is remarkable. It is part of His plan to reveal these things and to allow us to rediscover them. This is one of the joys found in science. One can uncover design and beauty in new ways from perspectives that have not already been utilized.

Another useful modality is the approach being embraced by the Intelligent Design movement. Using this approach, the scientific method is used, but with the methodological perspective that Divine intervention is allowable in scientific research and explanations. Thus creation can have a designer. So as a result, Stephen Meyer can consider a question that would be difficult from a methodological naturalistic paradigm: where did genomic information come from? If one considers intelligent design as a methodological foundation one can then design experiments based on this informed point of view and see where the data leads.

As a result, one can do experimental and theoretical research that seeks to determine whether fundamental concepts, such as natural selection, can be proven. For example, can Darwinian evolution transform a cell from one species to another? Can a protein realistically be changed from one protein to another via natural selection? Intelligent design proponents argue that there are significant explanatory challenges that Darwinian processes have to overcome. Two examples are the evolution of protein folding and enzyme origin. Determining an approach that explains protein conversions via mutations with reasonable probabilities is difficult. Reeves, Gauger, and Axe argue that, "The problem for evolutionary explanations is that the very special circumstances needed to achieve even weak [protein] conversions in the lab translate into highly unrealistic evolutionary scenarios" (Reeves, Gauger, and Axe, 2014, pg. 11). This forms an explanatory challenge that has not been resolved. Since a mechanism for transforming protein folds is difficult from a Darwinian process point of view, a more modest search for enzyme origins based on similarity and phylogeny is usually employed. The more modest search for enzyme origins using similarity and phylogeny is a less challenging process to establish with Darwinian methods, but so far the Darwinian methods have not established a mechanism. As noted by Reeves *et al.* below:

The greatest challenge facing evolutionary accounts of enzyme origins is explaining how enzymes with new fold structures first appeared. Having made the case that this challenge is insurmountable in Darwinian terms we turned our attention several years ago to the more modest challenge of explaining how enzymes that existed long ago might have been coaxed into putting their structures to new uses. Certainly there is no shortage of modern enzymes that use similar structures to perform different functions, and at first glance this may seem to fit the evolutionary account of enzyme origins. However, because the point of studying protein origins is to explain how the many different functions arose, a successful explanation of enzyme diversity will have to focus more on the differences than on the similarities. The fact that subtle structural differences among the members of enzyme families cause profound functional differences might suggest that these functional differences are easily achieved, but the accompanying sequence differences, which are substantial, could equally support the opposite conclusion.

(Reeves et al., 2014)

Poythress considers how scientific explanatory problems can be bounded and where theological insights might become relevant by considering what Eugene Wigner, a Nobel Prize winner in physics, examined in his works that ponder what the aggregated information might mean, such as Wigner's 1960 article, "The Unreasonable Effectiveness of Mathematics in the Natural Sciences." He saw the overwhelming usefulness of mathematics in natural science as an astounding fact, but considered it a mystery as to why this would be the case. Similar to Flew's conclusions, he notes that it seems strange that laws of nature would exist in the first place, and there is no rational explanation. Poythress points out the following:

Wigner is right to draw attention to something that practicing scientists usually take for granted, namely that there are such things as "laws of nature". The scientist takes it for granted. But why should he? Perhaps it is because his teachers before him also took it for granted. But that just pushes the question back into the past. Once Wigner starts asking questions, he finds himself amazed. The whole edifice of science has a foundation on prescientific trust—we might say prescientific faith in the existence of laws of nature. And, according to Wigner, it is "mysterious". It is not at all obvious why laws of nature even exist. And granted that they exist, why should they be of such a kind that human beings find themselves capable of discovering them?

(Poythress, 2014, pg. 16)

So, not only is methodological naturalism problematic during the application

Figure 6.3: Science and Theology Collaborative Method Emphasis in Research Timeline

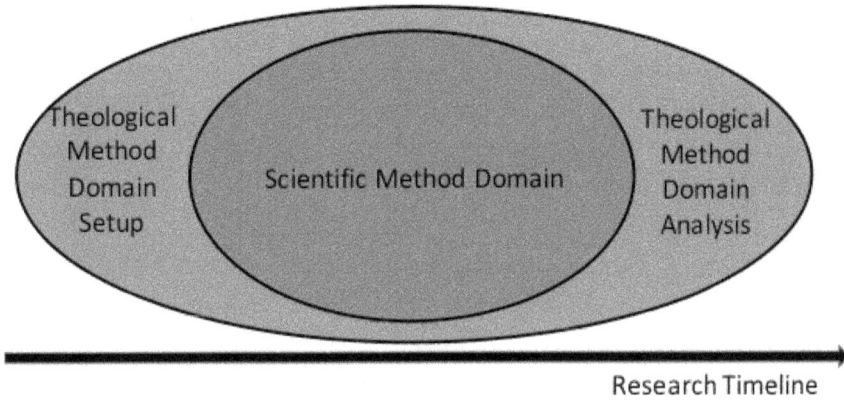

of the scientific method, the shortcomings of methodological naturalism occur even before the scientific method is even applied.

The theological method can be a tool to help guide scientific endeavors, set up and define experiments, and extend what can be understood by them. Consistent laws of nature are a natural outcome of a theological perspective. Figure 6.3 shows how the theological and scientific methods can vary in significance during the research timeline. In the beginning and towards the end, the theological method can have a greater impact. There is an answer to Wigner's dilemma that helps us understand the nature of our world and the characteristics of us who observe the world, as Poythress summarizes below:

> Thus, the Bible gives the explanation that Eugene Wigner was searching for and did not find. The laws of nature exist because God has spoken the universe into existence and continues to sustain it through His word of power: "He upholds the universe by the word of his power" (Hebrews 1:3, ESV). The Bible also explains why it is that human beings have an ability to discover laws of nature. It states that God made man "in the image of God" (Genesis 1:26–27, ESV). Human beings are creatures; they are finite. But they are very special creatures, whose existence reflects on a creaturely level some of the character of God. This reflection includes our ability to think. When we think, we imitate the original uncreated thinking ability of God.

(Poythress, 2014, pg. 18)

4 Problems with the Claim: What Others Assert by Limiting Scope

For many, science is viewed as the only source of truth and knowledge, and do not consider theological truth acceptable or allowable. Based on a naturalistic definition of science, any attempt at integration would not be looked at as science.

Alternative explanations exist and are viewed favorably in secular circles. Competing ideologies exist focusing on certain open scientific and philosophical arguments that may point to yet-to-be-determined mechanisms. In general, there are two approaches: one is to revise and augment hypotheses and adapt scientific theory to address current short-comings, and the other is to introduce philosophical arguments that may suggest a way of dealing with the discrepancies.

The approach of integrating theology and science can be used to better understand the source of genomic information. There are times when unsolved artifacts of the data frustrate researchers. Perhaps this implies there are aspects of its meaning that go beyond our understanding. With a Divine programmer and architect it is reasonable to assume His capability would go beyond ours and design approaches in the code would go beyond our comprehension, even for experienced researchers.

Adequately examining and validating the claims proposed in this chapter will take time and resources. The longer term strategy is to consider a real life genomic situation and apply both theological and scientific methodology in a structured manner. Philosophy has to be leveraged as a catalyst to bring the two domains together.

Appropriate engagements and boundaries between the two domains have to be defined. Where might theology be harmful to science? Where might science be harmful to theology? Carl Henry notes that trying to do theology with only scientific tools will lead to bad results, and is cautious when considering Lonergan's theological method approach of leveraging the scientific method in theology. Exploring theology by utilizing the scientific method brings the limitations of its approach into the study of God. Henry summarizes this way:

> Theology preserves its vitality only when it is sure of its own ground and engages in discourse with other disciplines. But on what basis will it pursue this dialogue? Rather than permitting theological subject matter to determine its own relevant method, Bernard J. F. Lonergan (*Method in Theology*) seeks in advance a method common to all sciences, theology included. He therefore risks the subsumption of some subject matter, especially that of theology, to other sciences. Lonergan affirms only the "virtually unconditioned" (hence highly probable, yet not beyond possibility of revision), and hence is indifferent to Scripture as a basic instrument of final truth. But a theological method derived from other sciences really adjusts theology to a general methodology that denies to theology its own distinctive object and subject matter. If the attempt to discuss

God's nature is confined to theological statements related to objects within the field of other objects, God will remain a mystery. Current religious knowledge-theory is historically conditioned by modern scientific controls and imposes upon theology an ideal borrowed from Leopold von Ranke's scientific historiography. Theology is not to be chained in advance to the method of other sciences. Does not a scholar like Lonergan, who can write almost interminably on epistemology and not mention God, need to reconsider the relation between revelation and reason? Nobody should be overwhelmed by a discovery that we cannot reach the Christian doctrines by Lonergan's method.

(Henry, 1999, pgs. 195–196)

5 Alternatives to the Claim: What Others Propose for Revised Explanations

Two alternatives categories are considered to the claim, and represent two perspectives that are active discussions in current literature. First, there are claims that a new element or capability in the natural order, that of innovation, emergence, or spontaneous self-organization exists. This is an interesting idea and can be self-consistent in its philosophical hypothesis, but is it true? How do we know if it actually exists? Second, there is a philosophical argument that we need to just look at the scientific facts differently to find a more compelling organization of the facts and better scientific arguments. James Shapiro and Stuart Kauffman are examples of the first case, and Thomas Nagel is an example of the second case.

Shapiro makes an argument that within natural genetic engineering there is the possibility for something new to come forth, or innovation. Somehow this could be manifested on its own with the right circumstances. He makes the following argument:

How does novelty arise in evolution? Innovation, not selection, is the critical issue in evolutionary change. Without variation and novelty, selection has nothing to act upon. So this book is dedicated to considering the many ways that living organisms actively change themselves. Uncovering the molecular mechanisms by which living organisms modify their genomes is a major accomplishment of late 20th Century molecular biology.

(Shapiro, 2011, pg. 1)

Shapiro comments on how living organisms undeniably have the capacity to alter their own heredity:

> The perceived need to reject supernatural intervention unfortunately led the pioneers of evolutionary theory to erect an a priori philosophical distinction between the "blind" processes of hereditary variation and all other adaptive functions. But the capacity to change is itself adaptive. Over time, conditions inevitably change, and the organisms that can best acquire novel inherited functions have the greatest potential to survive. The capacity of living organisms to alter their own heredity is undeniable.
>
> (Shapiro, 2011, pgs. 2–3)

These are interesting ideas and sound compelling, but how can one know whether they are true with only the information available within methodical naturalism? Meyer gives a rebuttal to Shapiro's arguments concerning innovation by commenting on the appearance of algorithmic complexity in an organism:

> ID [Intelligent Design] also makes predictions about the structure, organization, and functional logic of living systems. In 2005, University of Chicago bacterial geneticist James Shapiro (not an advocate of intelligent design) published a paper describing a regulatory system in the cell called the lac operon system. He showed that the system functions in accord with a clear functional logic that can be readily and accurately represented as an algorithm involving a series of if/then commands. Since algorithms and algorithmic logic are, in our experience, the products of intelligent agency, the theory of intelligent design might expect to find such logic evident in the operation of cellular regulatory and control systems. It also, therefore, expects that as other regulatory and control systems are discovered and elucidated in the cell, many of these also will manifest a logic that can be expressed in algorithmic form.
>
> (Meyer, 2010, pg. 484)

Stuart Kauffman introduces a similar idea to Shapiro by arguing for spontaneous self-organization as a new explanation for areas where natural selection is inadequate. He is trying to add a correction factor, if you will, to natural selection ideology. Thus he suggests spontaneous self-organization. At least he is open to talking about the short-comings of evolutionary theory, but he still is operating within the same methodological frame of reference. He tries to repair evolutionary theory:

> In my previous two books, I laid out some of the growing reasons to think that evolution was even richer than Darwin supposed. Modern evolutionary theory, based on Darwin's concept of descent with heritable variations

that are sifted by natural selection to retain the adaptive changes, has come to view selection as the sole source of order in biological organisms. But the snowflake's delicate sixfold symmetry tells us that order can arise without the benefit of natural selection. *Origins of Order* and *At Home in the Universe* give good grounds to think that much of the order in organisms, from the origin of life itself to the stunning order in the development of a newborn child from a fertilized egg, does not reflect selection alone. Instead, much of the order in organisms, I believe, is self-organized and spontaneous. Self-organization mingles with natural selection in barely understood ways to yield the magnificence of our teeming biosphere. We must, therefore, expand evolutionary theory.

(Kauffman, 2007, pg. 151)

From a philosophical point of view, Nagel makes an argument for teleological naturalism. He postulates a philosophical extension of science:

Pointing out their limits is a philosophical task, whoever engages in it, rather than part of the internal pursuit of science—though we can hope that if the limits are recognized, that may eventually lead to the discovery of new forms of scientific understanding. Scientists are well aware of how much they don't know, but this is a different kind of problem—not just of acknowledging the limits of what is actually understood but of trying to recognize what can and cannot in principle be understood by certain existing methods. My target is a comprehensive, speculative world picture that is reached by extrapolation from some of the discoveries of biology, chemistry, and physics—a particular naturalistic *Weltanschauung* that postulates a hierarchical relation among the subjects of those sciences, and the completeness in principle of an explanation of everything in the universe through their unification. Such a world view is not a necessary condition of the practice of any of those sciences, and its acceptance or nonacceptance would have no effect on most scientific research. For all I know, most practicing scientists may have no opinion about the overarching cosmological questions to which this materialist reductionism provides an answer. Their detailed research and substantive findings do not in general depend on or imply either that or any other answer to such questions. But among the scientists and philosophers who do express views about the natural order as a whole, reductive materialism is widely assumed to be the only serious possibility.

(Nagel, 2012, pg. 4)

Nagel argues that there is a bright future ahead of us for what may be discovered:

> It may be frustrating to acknowledge, but we are simply at the point in the history of human thought at which we find ourselves, and our successors will make discoveries and develop forms of understanding of which we have not dreamt. Humans are addicted to the hope for a final reckoning, but intellectual humility requires that we resist the temptation to assume that tools of the kind we now have are in principle sufficient to understand the universe as a whole.

(Nagel, 2012)

Nagle also acknowledges that the naturalist account is not satisfactory and other things should be considered:

> But for a long time I have found the materialist account of how we and our fellow organisms came to exist hard to believe, including the standard version of how the evolutionary process works. The more details we learn about the chemical basis of life and the intricacy of the genetic code, the more unbelievable the standard historical account becomes.

(Nagel, 2012, pg. 6)

Another well-known perspective that is held by some Christians is theistic evolution. The argument is that unguided process with right, Divinely-aligned initial conditions can produce life. Since this largely aligns with natural selection it is not dealt with in detail. One can argue that there has not been any proof, to date, that such initial conditions can be identified to support the complete evolutionary cycle of going from non-life to life, from simple life to complex life, and ultimately producing mankind who is made in the image of God.

There is no argument given for the source of innovation, self-organization, or emergence. It still has some of the same philosophical characteristics of natural selection. One might argue it is moving in the path towards theistic evolutionary ideals by suggesting that they are caused by creating the right conditions. Unfortunately, it cannot address the question of how those right conditions came about. Most likely they unwittingly could provide evidence of walking up the left side of the triangle towards science and theology integration as shown in Figure 6.2. Utilizing the integration triangle further, one might consider theistic evolution as a first step up from strict, unguided natural selection processes; a second step could be emergence coming about as the result of the right set of orderly conditions; and then a third step could be the intelligent design paradigm

The next section suggests an approach for when and how the theological method can be added in a structured way. This can address the areas where arguments like emergence theory are inadequate.

6   Filtering Analysis: An Approach That Includes Theological Method

In this section an approach is introduced that could be used to see if, when, and how theological method can be included in scientific exploration and data analysis. From a cross domain perspective, how can existing theories be evaluated? How can the gap between what is known and what is not known be quantified? The use of filtering (and effectively creating decision gates) is proposed. These clarifying filters are used in a step-by-step fashion to consider how theological and scientific methodologies could be applied together and provide more meaningful results. Filtering may give insight and may help quantify major components that are being analyzed in a controlled experiment. One of the goals is to see how one can quantify and utilize theological method in a structured way.

There are four levels of filters that can be used to support current evaluations of explanations, to assess when there is a need for added domain scope, to see how sensitive the major elements are in the data via sensitivity analysis, to support analyses of alternatives, and to evaluate the utility of experimental designs. This four level process is summarized below.

First the explanatory filter evaluates how well the current theory explains the experimental results. It does this in light of both scientific and theological sources of knowledge. It seeks to quantify what may not be explained adequately with the existing theory. It allows for both physical and nonphysical sources of knowledge. Second, if the first filter identifies a potential short-coming of the theory when only the scientific method is applied, then the next step is to determine whether the inclusion of the theological domain would help. What aspects of theological method are useful? For example, is there evidence of the character of God being shown in natural law that offers an approach worth considering further with specific lines of experimentation or analysis? Third, the sensitivity filter seeks to find what principle components seem the most promising and can be captured in an experimental design. This allows theology to be actively involved in the discovery process by helping one select a line of argument that might not be obvious if one only considered methodological naturalistic approach. Fourth, the explanatory filter seeks to quantify specific phenomena that supports a line of argument that has been identified at a high level in the third filtering step of sensitivity analysis.

1. Explanatory filter: how well does the current theory perform?

   - How well does the current approach explain the data?
   - What assumptions are required to make it valid?
   - What alternative theories exist?
   - What is not explained?

2. Domain filter: should the scientific method be augmented with theological method?

   - Can we determine if the theory is reality or just a well thought out hypothesis?

   - If things are not explained does the theological method help?

   - What possible sources of truth, knowledge, information, and insight could help?

   - How might this be useful?

     - Biomimicry: leverage nature design techniques in man-made products
     - "Theomimicry": this is coining a term to capture where we notice theological principles found in nature
       * Symphonic filtering methodology: Assessment of Scriptures that follow a particular theme

3. Sensitivity filter: what are the principle components to study?

   - Functional decomposition: what are the major elements to analyze?

   - Principle component analysis: what is the most significant element that should be focused on?

4. Characterization filter: what known signatures are present?

   - What "design pattern" or template concepts should be recommended for analysis and design of experiments?

     - Build a reference library over time of characteristics to consider

An additional useful feature in this dual domain approach is having a larger set of tools to objectively conduct analyses of scientific findings. This is another endeavor, analyzing experimental results to see if the findings make sense and if they correspond with reality based on the larger scope that is available by including theological and scientific methodologies. Understanding rich information sources like those found in genomes is a good example. This approach could provide justification for efforts to search the genome for higher-level non-structural elements like conformity to specific design pattern rules. In genomics this could be a focus area within bioinformatics or a tool used in sequence assessments. Examples of initial questions to evaluate scientific findings are listed below:

- Where did the information come from?

- What is the information content?

- How should the information be examined?

- What do results mean?

- What organizational principles can we make from the information analysis?

There are previously developed ideas that this approach is leveraging. First, there is the explanatory filtering assessment that William Dembski developed, described below. There is a multiple step process developed here that largely corresponds to the first level of filtering described above.

> As a criterion for detecting design, specified complexity enables us to decide which of these modes of explanation apply. It does that by answering three questions about the thing we are trying to explain: Is it contingent? Is it complex? Is it specified? By arranging these questions sequentially as decision nodes in a flowchart, we can represent specified complexity as a criterion for detecting design.
>
> (Dembski, 2004, pg. 87)

Second, there is the concept of inference to the best explanation, which is contained in the assessment process of the first two filters. This has been postulated by intelligent design proponents like Stephen Meyer. It seeks to establish the most compelling explanation of the experimental results. Meyer argues:

> For this reason, the design inference defended here does not constitute an argument from ignorance. Instead, it constitutes an "inference to the best explanation" based upon our best available knowledge.
>
> (Meyer, 2010, pg. 376)

Thus it is a way of considering all the possible explanations in a structured manner and of making a good decision based on that information. He also goes on to point out that intelligent design is testable, and, in fact, a body of research is being developed on this front:

> Critics of intelligent design often argue that the theory cannot be tested, because it makes no predictions. The charge turns on a fundamental misunderstanding of how historical scientific theories are tested. Primarily, such testing is accomplished by comparing the explanatory power of competing hypotheses against already known facts. The theory of intelligent design, like other theories about the causes of past events, is testable, and has been tested, in just this way. That said, the theory of intelligent design also has predictive consequences. Since the design hypothesis makes claims about what caused life to arise, it has implications for what life should look like. Moreover, the explanatory framework that intelligent design provides leads to new research questions, some of which suggest

specific predictions that are testable against observations or by laboratory experiments.

(Meyer, 2010, pg. 481)

7   Impact: Why Theological Method Can Make a Difference

Why does the claim matter? Coming to unjustifiable conclusions is a serious matter. As stated in the beginning of this chapter, if there are things that cannot be explained adequately by science, and if one cannot know the capability of one's measuring tools without an external calibration source or an absolute frame of reference, then theological insights and their knowledge claims are useful. The intent of an approach that is friendly to the theological method is to promote seeking and embracing of ultimate truth found with both theology and science.

It is a better use of resources to discover deeper truths rather than just a more refined self-consistent standalone theory that cannot be validated as correlating with reality. A reasonable argument for what genomic complexity research is showing us is that uncovering new experimental findings will require new methods to more fully understand reality. It may also require us to humbly accept that we cannot know it all.

The literature showed a few examples where published results do not align as well as the research community would like with the widely accepted methodological naturalistic tenets used as a basis for most science. Based on emerging results from several scholastic communities who are embracing the existence of factors that go beyond the physical, there is the opportunity to quantify the utility of theological method in practical ways. This potential can be seen by looking at the work of several individuals in these emerging communities mentioned in this chapter such as John Sanford, Doug Axe, and Ann Gauger.

Some may not consider embracing the theological method as scientific. In some regards one could argue this does not matter. If all truth comes from an ultimate source, maybe it is more appropriate to call it theology, or a more focused term more such as "applied theology." If the truths of cosmology and genomics lead us to Divine revelation this is a profound thing. The book of nature can reveal interesting bits of information that, in part, man is able to grasp. Reading this along with Divine revelation in a sense turns a black and white canvas into a full color masterpiece inspired by God awaiting us to ponder the aspects of His creation and his character as well as thinking his thoughts after Him.

A key part of this chapter is quantifying a structured process for how one can apply the theological method to scientific analysis and discovery. Future publications will show specific examples of applying these filters and seeing if this provides a more meaningful explanation of the results in the process.

References

Clark, D. 2010. *To Know and Love God: Method for Theology*. Crossway Books.

Dembski, W.A. 2004. *The Design Revolution: Answering The Toughest Questions About Intelligent Design*. InterVarsity Press.

Flew, A. 2008. *There is a God: How the World's Most Notorious Atheist Changed His Mind*. HarperOne.

Forterre, P. and Philippe, H. 1999. The last universal common ancestor (LUCA), simple or complex? *Biological Bulletin* 196:373–377.

Henry, C. 1999. *God, Revelation, and Authority*, volume 1. Crossway Books.

Kauffman, S. 2007. Prolegomenon to a general biology. In W.A. Dembski and M. Ruse (editors), *Debating Design: From Darwin to DNA*, pp. 151–172, Cambridge University Press.

Meyer, S. 2010. *Signature in the Cell: DNA and the Evidence for Intelligent Design*. HarperOne.

Nagel, T. 2012. *Mind and Cosmos: Why the Materialist Neo-Darwinian Conception of Nature Is Almost Certainly False*. Oxford University Press.

Nelson, P. 2016. Design triangulation. In *2016 Conference on Alternatives to Methodological Naturalism*, video.
https://www.youtube.com/watch?v=rNY_i1kJAnk

Poythress, V.S. 2014. Why is science possible? In D. Bundrick and S. Badger (editors), *Genesis and Genetics: Proceedings of the 2014 Faith and Science Conference*, pp. 15–22, Logion Press, Springfield, MI.

Reeves, M., Gauger, A., and Axe, D. 2014. Enzyme families–shared evolutionary history or shared design? a study of the GABA-Aminotransferase family. *BIO-Complexity* 2014(4).

Sanford, J. 2015. *Genetic Entropy*. FMS Publications.

Shapiro, J. 2011. *Evolution: A View from the 21st Century*. FT Press Science.

The ENCODE Project Consortium. 2007. Identification and analysis of functional elements in 1% of the human genome by the encode pilot project. *Nature* 447:799–816.

Wigner, E. 1960. The unreasonable effectiveness of mathematics in the natural sciences. *Communications in Pure and Applied Mathematics* 13(1).

Describable but Not Predictable: Mathematical Modeling and Non-Naturalistic Causation

Jonathan Bartlett

The Blyth Institute

Abstract

Our notions of causation in science are often unintentionally constrained by the mathematics we use. Typically, scientific investigations use algebraic or calculus-based mathematics to model causes and effects. In these types of models, there is a predictive relationship between the cause and the effect. This predictive pattern is what most people use to classify events as materialistic, leaving events that are not so classified as non-materialistic. Mathematics over the last century has introduced new formalisms that cover functions that do not conform to the materialistic pattern. While these functions cannot always predict outcomes for typical cases, they can be studied and analyzed in other ways, and therefore can be used for knowledge-building. Therefore, by expanding the mathematical toolset, investigators can better identify and model non-materialistic causes.

1 Introduction

One of the largest barriers to considering alternatives to naturalism is that it is difficult for people to conceive of what sorts of causes lie outside of the naturalistic paradigm and how they might be modeled. While some people can see that there must be more to the world than what is contained in naturalism, most view those parts of reality as fundamentally enigmatic. This has even led many who disagree with naturalism on a fundamental basis to do their work under a naturalistic rubric.

It has also prevented people from attempting rigorous studies of phenomena that are outside of naturalism's domain.

Therefore, for any alternative to methodological naturalism to take hold, methods of analyzing events that are beyond the reach of naturalism must be developed. Since non-naturalistic phenomena are categorically different than naturalistic phenomena, such methods will be necessarily different and may provide different kinds of information about the events than the kinds of information we are used to having. Such modeling should not be judged as successful or unsuccessful based on whether or not it matches the kind of information obtained from naturalistic models, but on whether or not it matches reality and provides helpful information that can inform decisions and can be combined with other forms of knowledge.

However, before we look at how to model non-naturalistic phenomena, we must first establish what it *means* for a particular phenomenon to be non-naturalistic.

2   Computability as a Demarcation

The first problem in developing alternatives to methodological naturalism is with determining what counts as a naturalistic or non-naturalistic phenomenon. While most people think that such a demarcation is intuitively obvious, on closer inspection, this becomes a rather difficult problem. Several attempts at creating a demarcation have been attempted and most of them have failed.

First of all, we will equate naturalism with physicalism—the idea that all knowable phenomena are in some sense physical. Without this restriction, naturalism just means "everything," and defining an idea as "everythingism" is unhelpful. Physicalism is what most people mean when they talk about naturalism. However, this leaves us to define what it means for something to be physical. Without a solid definition in place, the category is meaningless (Stoljar, 2009).

For instance, let us imagine that ghosts are real. Are they physical? By what criterion could we classify them either way? It is true that ghosts are often considered non-physical beings, but on what rule might we make such a determination?

Some might say that an event is physical if it is observable and detectable. Let's say, then, someone found a way to detect ghosts and their effects. Are ghosts then considered physical? A demarcation between physical and non-physical that puts ghosts on the "physical" side seems nonsensical. Others might say that an event is physical if it is testable. But, again, if we developed a mechanism that tests for ghosts, then ghosts would be physical.

Physicalists oftentimes define "physics" to include anything that has effects. However, non-physicalists believe that their categories (i.e., mind, spirit, etc.) do in fact have effects in the physical world. Therefore, not only is such a definition of physical nonsensical (given the ghost example), it also does not put itself at odds with the non-physicalist claims. Bartlett (2016) goes into more detail on the problems of demarcation for physicalism and materialism.

To make a real distinction between these classes of phenomena at all, we need an objective criterion that takes into account both the features that physicalists and non-physicalists find important. No criteria is likely to have universal agreement, but for a criteria to be worthwhile it should have at least some who agree with it on both sides of the issue.

One criteria proposed by physicalists to distinguish between physical and non-physical phenomena is computability. Under this rubric, physical processes are those whose results can, at least in principle, be calculated by computational systems, while non-physical processes are those which cannot. This thesis is described and defended more fully in Bartlett (2014).

This demarcation has many advantages. First, it is objective. Computation and computability is a well-studied topic. One can prove that certain mathematical functions are non-computational. Second, it is used by physicalists themselves. Having a demarcation criterion that is agreed upon by both parties makes discussion and progress possible. Third, since computation is a finitary mode of acting, having computation as a criteria means that events themselves must obey finitary logic, which is a reasonable requirement from a physicalist standpoint.

Therefore, for this paper, we will use computability as the demarcation criteria between naturalistic and non-naturalistic modes of causation.

3   Elementary Functions and Causation

Many of our ideas of causation, especially in the sciences, come from our experiences in mathematical education. I have no general criticism of mathematical education to offer (I myself teach math in the way described here), but it is important to note that the way math is taught and the way we learn it predisposes us to thinking about reality in certain ways. When lecturing calculus-level students, I try to at least alert students to the ways that their experience with lower-level mathematics biases them to thinking about the world in certain ways.

Most of the functions familiar to students are what I will call well-behaved functions. The biggest defining character of smooth functions, at least as I am considering them, is that they are smooth; specifically, they are C^ω smooth over a majority of the important parts of their range. C^k smoothness refers to the number of times that a function can be differentiated successfully resulting in a continuous function, with k indicating the number of times a function can be differentiated.

For instance, take the following function:

$$f(x) = \begin{cases} x^2 & \text{if } x \geq 0 \\ 0 & \text{if } x < 0 \end{cases} \tag{7.1}$$

This function has C^1 smoothness. The function itself is continuous, and its first derivative is continuous. However, its second derivative is discontinuous at $x = 0$, because the left-hand limit is zero while the right-hand limit is two.

Polynomials by themselves are all C^∞ smooth. Likewise, sin and cos are C^∞ smooth. Most functions that students encounter in the early years of mathematics are C^∞ smooth.

These functions are also C^ω smooth. This level of smoothness means that not only are the functions C^∞ smooth, but also that their Taylor series expansion converges. This is hugely important in the natural sciences for a quite unexpected reason.

If a function is C^ω smooth, then that means a small sample of the data can be extrapolated out infinitely simply by observing the small sample with more and more detail. For example, to do a Taylor series for a function f around a specific point A, I need only find the derivatives of the function *near* A in order to model the entire function f. As I find the value of higher and higher derivatives of f at point A, I get closer and closer to its behavior for any value at all.

The significance of this is that, for C^ω functions, to find the behavior of f at B, for *any* B, I need only to analyze the behavior of f at A in sufficient detail. I have no need whatsoever of analyzing f at B or even near B in order to have a reliable estimation of its value, assuming I can get enough resolution of the function f at A.

For a more specific example, let's say that I had a phenomenon P that followed the sin function such that the true effects of the phenomenon could be given by $P = \sin(q)$ (where q is some other quantity that is affecting the output), but I was not aware of the existence of the sin function, nor of the fact that P behaved that way. Let us also say that I could only vary q within the range of 0 to 0.8 with my experimental apparatus, but could measure both q and the result of the phenomena with infinite accuracy. Because the phenomena is C^ω smooth, this means that even with my limited data set, I can determine the behavior of P for *any* value of q using only this data.

So, if I vary q between 0 and 0.8 I will get the graph shown in Figure 7.1

But, if I measure the velocities, accelerations, accelerations of accelerations, and so forth, I can eventually get the full picture of what the phenomena is doing in the long term, as you can see in Figure 7.2.

Therefore, even though my sample is monotonically increasing and has a range between 0 and 0.0998 (and is in fact nearly linear in this range), if I were able to analyze it with sufficient detail, I could calculate its true range (-1 to 1), and determine that it is cyclical with a period of 2π. So, even though I analyzed a near-linear, tiny portion of the function, by being able to view its velocities, accelerations, accelerations of accelerations, and so on to a sufficiently high resolution, I can determine the large-scale patterns of behavior of the function.

For functions that work this way, it is a fantastic quality. The problem, however, is that all of the functions we work with in lower-level mathematics have this quality, and therefore students make the assumption that functions should and must work

Figure 7.1: The Sine Function Over a Restricted Domain

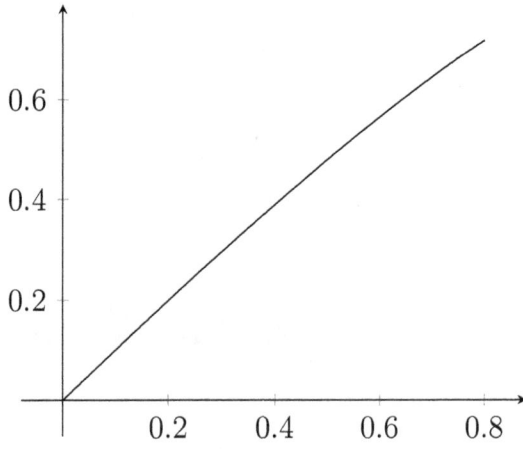

Figure 7.2: The Sine Function Over a Larger Domain

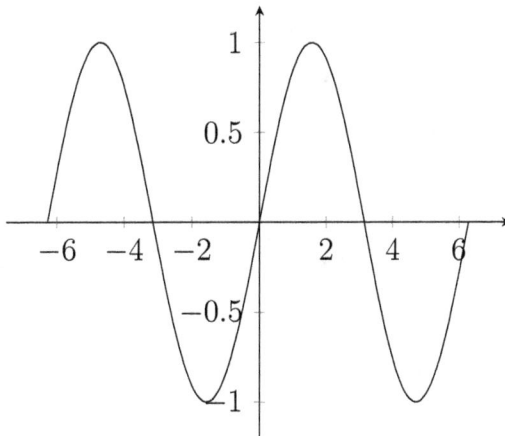

this way. The idea that there may be functions that do not behave this way is foreign to non-mathematicians. Therefore, when envisioning causation, it is assumed that causation itself must follow this pattern.

It may very well be true that causes that follow this pattern are easier to analyze using current scientific methodology. The question, though, is whether we want to limit science to only follow these causative patterns, or if we want to be able to find those that are more elusive.

As mathematical knowledge has progressed, more and more functions have been discovered that do not match traditional expectations in smoothness and other categories. However, their introduction into the sciences has been very slow due both to a lack of awareness and to metaphysical biases about whether or not causes can be modeled by them (van Rooij, 2008).

4   Ways to Be Strange: A Partial Survey of Strange Functions

In the previous section, we took a look at one way to be strange—by not being C^ω smooth. I should note that the purpose of the prior section was not to argue that all functions that are not C^ω are non-naturalistic, but rather to show how our expectations of how functions should behave intertwine with our expectations from nature, and how deviations from those expectation can affect both the way that we model nature and the way that we extrapolate our data to unknown values. In this section, we are going to look at what attributes of a function make it strange in a way that would lead us to consider that it is non-naturalistic.

As it happens, out of the total possible functions, most of them are not well-behaved. That doesn't mean that *realized* functions are mostly not well-behaved, but the number of available pathological functions greatly outnumber those that are well-behaved for almost any definition of pathological and well-behaved. Most of these functions at least appear useless and very likely are useless. However, there are several types of functions that are not well-behaved that may wind up being useful. Here we will do a short and incomplete survey of functions that may have the potential to model some interesting aspect of causation. We will term them "strange" rather than "pathological" to avoid the automatically negative connotations. The goal of this survey is not to say that any given function actually does model some event or process in real life, but to stimulate the imagination to help people think outside of traditional ways of thinking about causation and especially about *modeling* causation.

Figure 7.3: The $\omega(n)$ function from 0 to 100

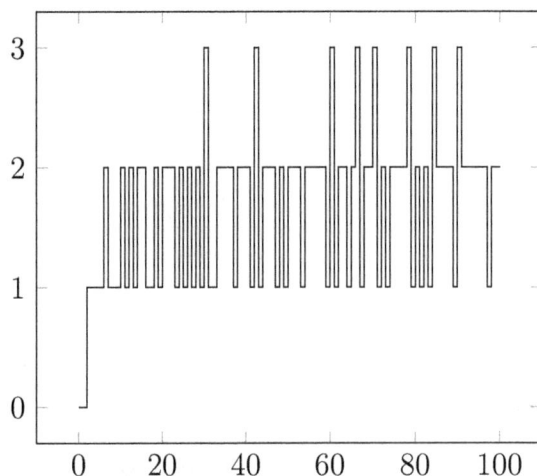

4.1 Primes, Number-Theoretic Functions, and Lengths of Causality

Number theory has some of the easier-to-understand strange functions. Many of the functions in number theory are discrete rather than continuous so the question of smoothness doesn't really apply, though you can use analytic continuations to create real versions of some of them.

A simple example of a function from number theory is the prime number function, where Prime(n) yields the nth prime number (starting with 2). For instance, Prime(5) would yield 11 because it is the 5th prime number.

Another interesting function from number theory is the distinct prime factors function, $\omega(n)$. (This is a different ω than the one for smoothness.) For this function, the result is the number of *distinct* prime factors for the given number. Therefore, $\omega(64)$ yields 1, because 2 is the only prime factor, but $\omega(30)$ yields 3 because it has prime factors 2, 3, and 5. Figure 7.3 shows a graph of this function.

What makes prime-oriented number-theoretic functions interesting is that the "length" of "causation" for each value varies. That is, the prime number 2 exhibits a causal relationship to every other number in the number chain, but the prime number 3 only exhibits a causal relationship to every third number in the chain. A prime number, then, is one in which there is no previous "cause" in play for its results.

Thus, while we normally think of causes as having uniform influences on future events as other causes, number-theoretic functions help us to think about non-uniform causation, where the reach of individual causes varies as well as the number of causative influences of any given event.

Figure 7.4: The Cantor Function on the Unit Interval

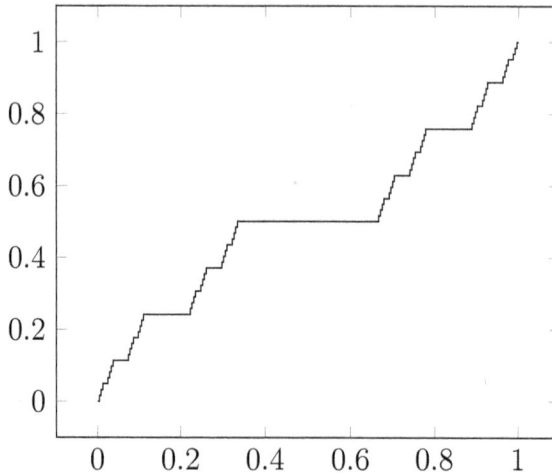

4.2   The Cantor Function and Two-Way Causation

A classic strange function is the Cantor Function. This function is strange because it is continuous over real numbers, almost entirely flat (i.e., non-flat spaces occur over a measure zero of the function space), and its y value increases from 0 to 1 over the space of 0 to 1 on the x-axis. Wherever it is not flat, it has an undefined (i.e., infinite) derivative. The graph is given in Figure 7.4.

The graph of the Cantor function can be constructed as follows:

1. Start with a left-hand point at $(0, 0)$ and a right-hand point at $(1, 1)$.

2. Find the point exactly halfway between the left-hand point and the right-hand point, which we will call the midpoint. On the first iteration, this will be $(\frac{1}{2}, \frac{1}{2})$.

3. Draw a horizontal line $\frac{1}{3}$ of the horizontal distance between the left-hand point and the right-hand point, centered on the midpoint, which we will call the midline.

4. Perform steps 2–4 again twice. The first time we will use the current left-hand point as the new left-hand point and the leftmost point of the midline as the new right-hand point. The second time we will use the current right-hand point as the new right-hand point and the rightmost point of the midline as the new right-hand point.

This process will repeat indefinitely, with more and more intervals being created each time. Assuming it proceeds an infinite number of times, it will be a continuous function as described above.

What is interesting about the Cantor Function is that it describes two-way causation. That is, the function starts with left and right points, and the middle points are decided based on collaboration between them. Since many of the concepts of alternatives to methodological naturalism involve some sort of dualism between mechanical causes (where the initial conditions determine the outcome) and teleological causes (where the final desired outcomes determine the intermediary steps), this seems to be a place where a Cantor-like function may be beneficial.

Cantor-like functions establish a way to associate feedback between mechanical causes on the left-hand side and teleological causes on the right-hand side. While the Cantor Function itself may or may not model anything in particular in real life, this *type* of function can help us expand our reasoning capabilities to be able to model multidirectional causation.

4.3   Assertion-Satisfying Functions

In computer programming, Landin (1965) developed the concept of *continuations*, which is an abstract control structure, where your location in the program can be "bookmarked" to return to later. This allowed for the development of a variety of interesting control structures that work quite differently than the typical ones considered in computer programming (sequence, selection, and iteration).

One of the more interesting ones is the ability to develop assertion-satisfying functions. In an assertion-satisfying function, variables can take on ambiguous values (i.e., the variable q could be assigned any of the values between one and five). Then, during execution, a particular value is tried, but if the results do not satisfy a downstream assertion, the results get thrown away, the function is backed up to the bookmark, and a new value for q is tried. It is possible that reality works like this. It is possible that some of the laws of nature exist as assertions and that reality may be able to back up and "try again" if the assertions do not match. In such a scenario the laws of physics wouldn't so much dictate what must happen, but rather only the constraints of what could happen.

4.4   The Halting Problem and Related Functions

Another type of function that should be considered in modeling reality is the halting problem and related functions. This is described more fully in Bartlett (2014), but a condensed version will be provided here.

In computer programming, a computation is supposed to finish—it is supposed to complete and yield a result. This is known as "halting." Computations that do not finish are said to be caught in an "infinite loop." If you have ever had your computer stuck in a situation where the cursor just spins and spins and never stops, it is possible you have experienced an infinite loop. Computations are supposed to yield values, and when they don't, it leads to problems.

The deeper issue, however, is determining if a computation will halt or if it will get stuck in some sort of infinite loop. One of the first discoveries of computer science was the fact that there is no algorithmic way to tell if a given computation will halt or if it will go into an infinite loop. That is, I cannot write a program that will tell me, even if I know all of the inputs, whether or not a different program will yield a result or go into an infinite loop. This is known as the "halting problem" in computer science. For any given program/input combination, it will either halt or it won't—there is no other possibility. However, figuring out which ones won't halt is an impossible problem for a Turing-like machine.

There are many problems in computer science that are essentially incomputable but could be computed if we were counterfactually able to write a program that solved the halting problem. One such problem is the "busy beaver" problem. In the busy beaver problem, the goal is to find, for a program of length N, what the largest output a program can generate and still halt is. This is generally unsolvable, but could be easily solved if we were able to write a function to determine the result of the halting problem.

Even though we can't implement a function to tell us whether a function halts or not, we can reason about how such a function can be used and what sorts of properties it would have. An unimplementable function is called a Turing *oracle* or just an *oracle*. Thus even though we couldn't *implement* or *predict* the results of an oracle, those are not the only types of reasoning available. Alan Turing, for instance, used oracles to measure the relative difficulty of different types of problems, showing that some problems are more or less complex based on the kind of oracle required. Therefore, just because a function can't be explained in terms of its operation, this doesn't exclude it from useful reasoning or knowledge-building.

Additionally, as argued in Bartlett (2014), it appears that humans are able to solve something similar to the halting problem. If humans could not tell if a program would halt or not, they would not be able to successfully program computers. Therefore, it seems that humans have access to an oracle of some kind that allows them to solve problems that are beyond computation. Robertson (1999) points out that the development of mathematical axioms is itself a super-computational problem. Therefore, the ability of humans to develop mathematics itself shows that humans have access to some sort of Turing oracle.

Bartlett (2014) suggested that perhaps the oracle humans have access to is the ability to generate needed axioms based on existing problems. The oracle was described as $A = I(Q, p, i, B)$, where Q is the problem the human is attempting to solve (with inputs p and i), and I is the human "insight" oracle function that reveals the set of axioms, A, needed to solve the problem Q. The function requires that the human already have B—the set of all axioms needed to solve the problem, except one. What is being proposed by the oracle function is that human insight is able to generate axioms (a non-computational event as described by Robertson (1999)) when humans are given a problem they cannot solve and all of the axioms they need to solve it, except one. Thus, even though this function isn't computable, it can be used to reason about non-mechanical models of the mind.

5 Using Incomputable Functions in Modeling

As we have seen, by expanding our view of mathematics beyond the typical well-behaved functions, we can incorporate models of non-mechanical (i.e., non-naturalistic) causes and modes of operation into mathematical descriptions. Such mathematical descriptions would enable better integration between naturalistic and non-naturalistic causes at work in a system. For a system to be mathematical does not mean that it must be predictive or even computational. Many functions in mathematics are not computable, and in fact, computability may only cover a very small proportion of them.

One may wonder, why bother with mathematics at all? Mathematics is just a formalization of logic. Using mathematics requires that a person distill their ideas into the most rigorous and abstract form. Additionally, because mathematics as a discipline is well-studied, boiling ideas down into mathematical forms, even if incalculable, allows mathematical tools to be used to analyze and reason from these ideas. It also makes it easier to combine different ideas. If each idea is expressed mathematically, then the combination of ideas can likely be expressed mathematically, and the logical consequences of these ideas can be more readily determined.

One of the advantages of our well-behaved functions is that their long-term behavior can be arbitrarily extrapolated from limited observations. Additionally, with a finite set of observations, it is difficult to distinguish a strange function from a more well-behaved nearby function, especially within the limited data set. A major issue with using strange functions in modeling is determining whether or not such a function is what is being observed or if its more well-behaved nearby function is at work.

Therefore, what is required for establishing a strange function as the basis for an observed effect is a *logical* reason for preferring the function. That is, there must be something in the nature of the causal relationships that would indicate the usage of a strange function in a model. One other thing that may indicate that a strange function is indeed required is the need to continually change the model with more and more data. This may indicate that a strange function is at play, and the causal relationships should be investigated to see if a strange function may properly model what is happening.

Since this paper only presents a small smattering of the known strange functions, more and more modeling power will be available by understanding more and more strange functions and how they relate to causality. The ones presented here were picked because the author could see ways in which they may be important to helping researchers think about causality. Additionally, more functions may be available by simple creative construction. Once the strictures of well-behaved functions are removed, and a person gets acquainted with the nature of strange functions, the ability to construct definitions of new strange functions to match the causality in question will be increased.

While strange functions should not be introduced lightly, there is no reason to avoid them in models. The preference for well-behaved functions is just that—a preference. There is no reason why reality must conform to our preferences. Linear components in models are better-behaved than non-linear components, but that doesn't mean our models must always contain linear components. The goal of modeling for science is to provide a deeper understanding of the nature of the subject under investigation. Other goals (such as using models for engineering) may substitute similar well-behaved functions in order to simplify calculations, but science, as an attempt to learn more about the true nature of reality, should in most cases prefer whatever function is the truest model of reality.

6 Testing Models that Use Strange Functions

One of the key features of science is testability. As pointed out in Bartlett (2016), the two main features of methodological naturalism that made it successful were that it defined a scope of inquiry and it provided a system of justification. With well-behaved functions, the system of justification is fairly straightforward. The model will predict how a system will behave for tests that have not been performed yet; the experimenter will then perform the tests and see if the results match the model within a margin of error. With strange functions, however, the models do not always predict behavior. Therefore, the system of justification used will have to be modified in order to accommodate strange functions.

However, before we look at how we can test our new models, we should think about why it is that we test models in the first place. The goal of testing is to allow reality to push back on our ideas. That is, we have ideas about the nature of reality, but our ideas must *conform* to external reality, not the other way around. Testing is done to make sure that reality has a chance to give us feedback on the truth of our ideas.

We should recognize that testing is not an absolute truth-teller. It is more of a sanity check than a rigid determiner of truth. For a finite set of data points, there are infinitely many functions that would be within the margin of error for those data points (Kukla, 1996). So how might someone decide between two empirically-equivalent theories? As it stands, our scientific ideas do not emanate entirely from empiricism. If they did, then this would be a problem. Instead, empiricism provides the dataset that we use to establish rationalistic models, and it provides additional data to validate such models. But the models themselves are based on logical relationships between entities under investigation.

Therefore, the key to testing is not that the data points must be uniquely determined by the theory, but rather that the theory must flow from a proposed logical relationship between entities and the data must be consistent with it within a margin of error.

Even though strange functions are not always calculable or predictive, they do lend themselves to reasoning about relationships, otherwise they would not be considered functions. Therefore, it is possible to find patterns that are true with a strange function that may be tested for, even if the strange function itself is not directly testable. In the next section, we will look at a specific example.

7   Randomness as an Exemplar Strange Function

While strange functions generally have not been given much scientific weight, one in particular has been used regularly—randomness. While randomness is not well-behaved like most of the functions within science, the willingness of scientists operating under methodological naturalism to use it probably stems from the fact that it does not appear on the surface to imply teleology (though see Bartlett (2008) for an alternative view). Therefore, its adoption in the scientific community as a viable model allows us to demonstrate the utility of strange functions in scientific modeling.

Randomness is actually a property of an infinite sequence. Therefore, no finite sequence of events can prove that the sequence is random. Additionally, randomness is not predictive. Therefore, including randomness in the model does not help to predict any specific outcome. Thus, randomness matches what we have called strange functions.

Since randomness doesn't predict a specific outcome and cannot be tested directly, how was it included in scientific theories? Basically, if a model of an event has multiple possible outcomes, and the outcomes proceed in an order that is statistically stable but does not point toward any other structure, then the suggestion of randomness is quite appropriate.

For instance, in a Poisson distribution, the mean is equal to the variance. Therefore, one can "test" for such a distribution by checking the mean against the variance. If they are close, then the suggestion that the process is a random process following a Poisson distribution can be maintained. There are many different ways that the mean can equal the variance, but if our formal reasoning leads us to expect such a distribution, and the distribution's characteristic features match our expectations, then the test can be considered confirmed. This is used, for instance, in the Luria-Delbrück experiment where the test for randomness is used to determine if a mutation is in *response* to a selective pressure or if the mutation preceded the selection.

8   Conclusion

The goal of the present paper is not to propose a specific idea or procedure, but rather to assist researchers proceeding in directions at odds with methodological naturalism by pointing to the stranger aspects of mathematics that can serve as tools when investigating non-naturalistic phenomena. My hope is that researchers will be able to unshackle their imaginations from the mathematics of naturalism but without losing the rigor necessary to develop well-founded theories of how different aspects of the world works. Likewise, making use of mathematics even when it isn't well-behaved will better enable integration of different models and phenomena.

References

Bartlett, J. 2008. Statistical and philosophical notions of randomness in creation biology. *Creation Research Society Quarterly* 45:91–99.

Bartlett, J. 2014. Using Turing oracles in cognitive models of problem-solving. In J. Bartlett, D. Halsmer, and M.R. Hall (editors), *Engineering and the Ultimate*, pp. 99–122, Blyth Institute Press, Broken Arrow, OK.

Bartlett, J. 2016. Philosophical shortcomings of methodological naturalism and the path forward. In J. Bartlett and E. Holloway (editors), *Naturalism and Its Alternatives in Scientific Methodologies*, Blyth Institue Press, Broken Arrow.

Kukla, A. 1996. Does every theory have empirically equivalent rivals? *Erkenntnis* 44:137–166.

Landin, P.J. 1965. A correspondence between algol 60 and church's lambda-notation: Part i. *Communications of the ACM* 8(2).

Robertson, D.S. 1999. Algorithmic information theory, free will, and the Turing test. *Complexity* 4(3):25–34.
http://cires.colorado.edu/~doug/philosophy/info8.pdf

Stoljar, D. 2009. Physicalism. In E. Zalta (editor), *The Stanford Encyclopedia of Philosophy*, The Metaphysics Research Lab, fall 2009 edition.
http://plato.stanford.edu/archives/fall2009/entries/physicalism/

van Rooij, I. 2008. The tractable cognition thesis. *Cognitive Science: A Multidisciplinary Journal* 32(6).
http://staff.science.uva.nl/~szymanik/papers/TractableCognition.pdf

8 || Methodological Naturalism and Its Creation Story

ARMINIUS MIGNEA

Abstract

The objective of this paper is to survey how science, when aligned to materialist philosophy through methodological naturalism, answers questions about the origins of some of the entities that present the highest interest to science. It will discuss the elements of the materialist origins narrative and how successful these elements are in providing support for the exceptional claim of materialism: that everything (meaning among other things, life, living organisms of all sorts, brains, our solar system, our Earth, minds and consciousness) "can be explained as manifestation or result of matter." The paper analyzes the nature and the internal complexity of living organisms through the use of the concept of *machines*.

1 Introduction

One succinct characterization of Methodological Naturalism (MN) is proposed by Barbara Forrest (quoting Paul Kurtz):

> [**Methodological Naturalism**] is committed to a methodological principle within the context of scientific inquiry; i.e., all hypotheses and events are to be explained and tested by reference to natural causes and events. To introduce a supernatural or transcendental cause within science is to depart from naturalistic explanations.

> (Forrest, 2000)

A more succinct characterization of MN is captured below:

> Methodological Naturalism is a strategy for studying the world, by which scientists choose not to consider supernatural causes—even as a remote possibility.

(methodological naturalism, 2016)

Methodological Naturalism is founded on the materialist philosophy. **Materialist philosophy** is well summarized as follows:

> A theory that physical matter is the only or fundamental reality and that all being and processes and phenomena can be explained as manifestations or results of matter.

(materialism, 2016)

This study is going to focus on some of the entities that present the highest degree of interest for science and provide particular challenges for materialism and methodological naturalism to explain their nature and origins. The entities to which we are going to give special attention in our study are: life in general; living organisms and humans (*Homo sapiens*) in particular; brains, minds, natural human intelligence, and consciousness; as well as our Earth and our Solar System, which provide the cradle for life.

Before we proceed to understand how materialism and methodological naturalism answer the origins question for the objects of interest, we will have an in-depth look at the natures, characters, degrees of complexity, and intricacies of the objects of interest enumerated above. If an object is simple then the task of providing a credible origin explanation may not be so challenging. On the other hand if the object under study has a complex structure, a high level of internal organization, and intricate inner workings, then the task of articulating a coherent origin story can be proportionally challenging.

2   Machines and Living Organisms

We will use the concept of machine to develop an understanding of the nature and degree of intricacy of our objects of interest. A machine is an easy-to-understand concept since we are familiar with various types of machines that we encounter in our life. Also, by using the concept of machines, this will provide an empirical and materialist-sympathetic approach for our inquiry.[1]

---

[1]There are alternative, non-machine views of the organism, but since those are not as sympathetic to the viewpoint of materialism, they will not be considered in this study.

2.1   A Short Theory of Machines

A machine is usually defined as a material object made of multiple parts that produces a useful function. We will use a more formal definition:

> A **machine** is an assemblage of components that have well-matched interfaces and interactions that accomplish together a sustained or repeatable function.

An example of a simple machine is a pair of scissors whose function is to cut sheet-like materials. It has basically three components: two blades articulated by a screw (see Figure 8.1). There is a thumb blade and a ring blade shaped such that the thumb or the ring finger control the action of the blades that have facing sharp edges. Although scissors are simple machines made only of three components, we know that there are many elements and conditions that make good scissors. Among them are good material (e.g., steel for all parts), properly shaped blades with sharp edges, properly sized and smooth-surfaced rings, sharp and resistant edges, hollow grind, properly sized and matched blades, and properly adjusted screws tightened to ensure proper mobility of the blades.

It is important to observe that even for one of the simplest machines, we can develop a full theory on how to craft a good quality pair of scissors that reliably executes its sustained function, cutting or shearing certain materials.

Machines cover a wide spectrum of complexity or number of parts. We mention in some order of increasing complexity: a windmill, a bicycle, a car, an airplane, a computer. Complex machines can be composed of sub-machines with well-defined roles and well-matched interfaces and interactions between the sub-machines.

There is an empirical corollary related to the concept of machine as defined here:

> Wherever there is sustained function there is a machine.

2.2   Living Organisms Seen as Machines

In the rest of this essay, we are going to look at living organisms as machines. Why? One of the main reasons is to facilitate the understanding and study of living organisms. Secondly, because knowing that living organisms have many rich functions, it is natural to investigate and discover the machines within the living organisms that deliver these functions. A third reason is that the machine metaphor is probably the best metaphor that can be used to investigate the parts and the inner workings of these organisms (especially under materialism). The machine is a concept and metaphor that is familiar to both materialists and non-materialists and can capture or model a wide spectrum of complexity. The routine investigations of the life scientist are usually motivated by a desire to identify the machinery that drives relevant biological processes such as blood circulation, blood coagulation or photosynthesis.

Scissor Terminology

Honed Scissor

Tip ....

Blade ....

Back ....

Ring Blade or
Thumb Blade ....

Inside Ring ....

Hone Line

Hollow Grind

Screw or
Adjusting Screw

Wing

Nail Hole

Ride

Hook Blade or
Finger Blade

Measured As:
USA - Tip to Ring
Europe - Tip to Rest
Asia - Tip to Ring

Bumper or
Silencer

Finger Pin,
Tang or
Finger Rest

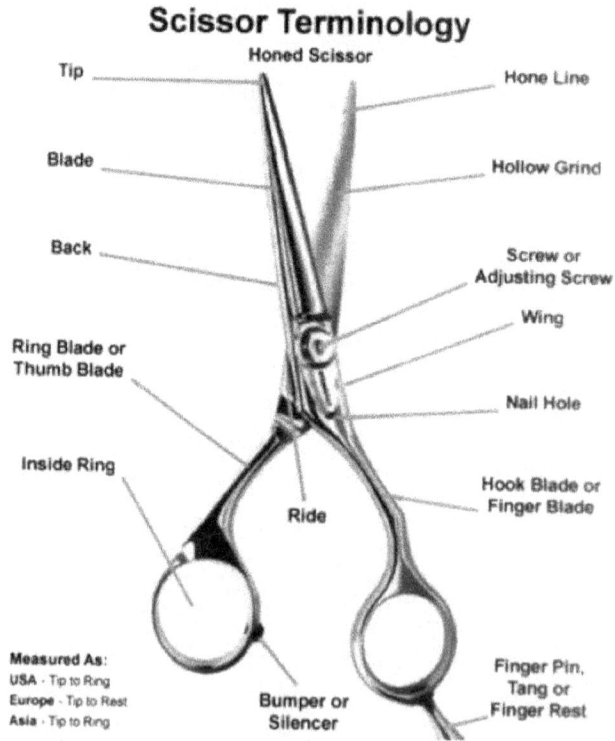

Figure 8.1: Scissors as an example of a simple, three-part machine

Photo credit: Image Copyright © 2009–2016 Precision Sharpening, Inc.

The analysis given in the following sections leads to the conclusion that living organisms can legitimately be considered super machines. A first, significant step of this analysis is performing a systematic comparison between Man-Made Machines (MMMs) and Living Organism Machines (LOMs).

3 Comparing Man-Made Machines (MMMs) with Living Organisms Machines (LOMs)

The comparison between MMMs and LOMs is being made from multiple perspectives:

- Comparing the number, nature and sophistication of functions

- Comparing the number of components and number of integrated subsystems

- Comparing the level of autonomy and self-sufficiency

- Comparing the degree of reciprocal integration and integration with nature

- Comparing body structure and body nature

Besides these comparisons, there are asymmetrical comparisons between the two categories of machines. Only Living Organism Machines manifests the following capabilities:

- The ability to self-replicate

- The ability to self-heal and recover after illness or after some body injuries

- (Specific to humans) The ability to reason, think abstractly, problem solve, and in particular, to imagine, design and implement machines

3.1 Comparing the Number, the Nature and the Sophistication of Functions

Man-made machines (MMMs) come in a wide variety of sizes and complexities. In general, even the most complex and sophisticated MMMs have a main function and a series of lower-level, secondary, or supporting functions.

If we consider a modern car, its main function is to transport people. There are also secondary functions, such as shelter and passengers' comfort (e.g., air conditioning, radio, or CD/DVD player), or navigation assistance with a GPS. The main function is provided by the internal combustion engine, while secondary and supporting functions are provided by various submachines: the transmission, the steering system, the fuel injection, the alternator, the fuel pump, the compressor, the electrical wiring, and the lights. The modern car is a sophisticated machine that achieved a higher level of sophistication with the introduction of various electronic and semi-conductor elements such as microcontrollers, on-board computers, and software that guides and controls the functioning of many of the car's sub-systems.

If we now consider a cat as a Living Organism Machine, we can distinguish, first, some functions on the lower-level of a functional hierarchy. Lower-level functions include the gastro-intestinal system with the cellular support for food metabolism and energy generation, the circulatory system, the pulmonary system, the musculoskeletal system.

On top of these lower-level functions, we can distinguish higher-level functions that are detectable outside of the body. In this category we have the cat's ability to eat food and drink liquids and to eliminate metabolic waste; cat's mobility and agility, including walking, running, jumping, catching, scratching, and biting; the cat's ability to perceive her environment concurrently on multiple dimensions or channels, including visual, audio, olfactory, taste, and touch.

On an even higher functional level, a cat manifests very sophisticated functions that are fundamental to self-preservation strategies that dynamically engage and choreograph the lower-level functions to achieve various self-supporting objectives, such as searching for and finding food, searching for and finding shelter, hunting and catching prey, hiding, defending itself, fighting, and searching for and finding mates with whom to sexually reproduce and rear kittens.

Although modern MMMs, such as a car or a fighter plane, have a good number of secondary or supporting functions besides their main function, the functions of the cat LOM are several classes of complexity beyond the functions of modern day MMMs.

This enormous complexity gap between the nature and number of functions of MMMs and LOMs is powerfully evidenced when we compare somewhat similar machines: a robot MMM and the human LOM, which serves as the model for the robot machinery. When we compare the bodies and physical abilities of a robot versus a human, some robots have more strength than humans, but the human has a much larger repertoire of movements and capabilities than robots do. If we compare a robot's abilities to sense its environment with a human's, it has only primitive abilities in terms of seeing or making sense of the sounds, speech, or sights around it.

Some very advanced robots may have a low-level degree of intelligence—i.e., artificial intelligence—in terms of making sense visually of their environment and moving and planning their actions accordingly. However, there is no comparison to the natural intelligence of humans that have superior environmental sensing and perception capabilities, memory, problem-solving capabilities, and abstract thinking functional capabilities (Bartlett, 2014; Thiel and Masters, 2014; Fodor, 2001). In particular humans' natural intelligence is what makes them capable of imagining, designing, and constructing machines (MMMs).

3.2 Comparing the Number of Components and the Number of Integrated Sub-Systems

In this section we will present side-by-side comparative counts of the estimated number of parts and components of MMMs and LOMs. For a car or an airplane, there is information available on the number of parts, mechanical, electrical, or of a different nature, that make up the machine. For a single-cell organism, the parts considered are molecular components that make up the organism, such as Messenger Ribonucleic Acid (mRNA), ribosomes, DNA base pairs, proteins, or membrane lipids.

The mapping between the MMM specified in the left column to the LOM specified in the right column is rather arbitrary. It was selected only for illustrative purposes.

When we compare a car MMM number of parts (ignoring the number of semiconductor/transistor components in a modern car) with the yeast (*saccharomyces cerevisiae*) we have about 30,000 parts for a Toyota car and counts of 120,000,000 DNA base pairs, 50,000,000 proteins and 1,000,000,000 membrane lipids for yeast, a single-celled organism.

The count of components in the LOM column is significantly larger when we do not consider semiconductor elements and their extremely large number of transistors in modern MMMs.

When we consider the number of transistors in the semiconductor components used by a hypothetical modern commercial airplane about the size of a Boeing 777, the number of components in such MMMs is dominated by the number of transistors in the DRAM memory chips, which is estimated to be $1000 \times 137,438,953,472 \simeq 10^3 \times 10^{11} = 10^{14}$.

Additionally, since we could not find any figures about the number of parts in an aircraft carrier or in the Hadron Super Collider (HSC) in Switzerland—considered one of the most complex machines in the world—we used our own common sense estimate for both. Each has 1000x more parts then the hypothetical modern commercial airplane. Thus our estimate for the number of parts in an aircraft carrier or in the HSC is $10^{14} \times 10^3 = 10^{17}$.

In the right column, when we estimated the number of components in the human body, we used the current estimate of the number of cells in the human body—37,000,000,000,000—and multiplied it by the number of components in the HeLa cell—1,000,000,000 proteins—to compute an overall component count for the human body as $37 \times 10^{12} \times 10^9 \simeq 10 \times 10^{12} \times 10^9 = 10^{22}$.

We observe, then, that the human body has $10^{22}/10^{14} = 10^8 = 100,000,000$x more components than the hypothetical modern commercial airplane. Using a similar computation we find that the human body has $10^{22}/10^{17} = 10^5 = 100,000$x more components than an aircraft carrier or the Hadron Super Collider.

It is interesting to note that the human body has 100 million times more parts than a commercial airplane even though it weighs about 3500 times less than the

Figure 8.2: Comparison of Component Counts of Man-Made
Machines (MMM) with Living Organism Machines (LOM)[a][b]

Man-Made Machine (MMM) component counts	Living Organisms Machine (LOM) component counts
Example MMM: *Toyota Car*: 30,000 parts	**Example LOM:** *Yeast (single cell)*: 30,000 mRNAs, 100,000 ribosomes, 10,000,000 membrane proteins, 120,000,000 DNA base pairs, 50,000,000 proteins, 1,000,000,000 membrane lipids.
Example MMM: *Boeing 777 Airplane*: 3,000,000 parts	**Example LOM:** *Mammalian HeLa cell*: 200,000 mRNAs, 1,000,000 ribosomes, 1,000,000,000 membrane proteins, 3,000,000,000 DNA base pairs, 1,000,000,000 proteins, 10,000,000,000 membrane lipids
Example MMM: *Intel 15-core Xeon Ivy Bridge-EX microprocessor* 4,310,000,000 transistors	**Example LOM:** *carnationpinkHuman Body*: Has $\simeq$ 37,000,000,000,000 cells $\simeq 10^{13}$ cells. A human cell (see HeLa above) has 1,000,000,000 proteins, that is: 10^9 proteins (parts).
Example MMM: *Xilinx Virtex-Ultrascale XCVU440 Field Programmable Gate Array (FPGA)*	So the human body has: $10^{13} \cdot 10^9 = 10^{22}$ parts (proteins)
Example MMM: *Samsung 128 Gb DRAM Memory* 137,438,953,472 parts $\simeq$ 10^{11} parts	**human body vs. commercial airplane** * Ratio: human body/airplane $= 10^{22}/10^{14} = 10^8 \times$ 20,000,000,000+ parts. * 100 million more parts in the human body. * Weight ratio: human body/airplane $= 10^2 \,\mathrm{Kg}/35 \times 10^4 \,\mathrm{Kg} = 1/3500$. **Human Body versus Aircraft Carrier or Hadron Super Collider**
Example MMM: *Super, modern commercial airplane with 1000 Samsung 128 GB memory chips* [generous estimation] Total: $10^{11} \times 10^3 = 10^{14}$ parts	* No. of parts in aircraft carrier: 10^{17} * No. of parts (proteins) in the human body: 10^{22} * Ratio: human/body/aircraft carrier $= 10^{22}/10^{17} = 10^5$ * 100,000 more parts in the human body than in an aircraft carrier OR in the Hadron Super Collider
Example MMM: *Aircraft Carrier (or Hadron Super Collider)* [generous estimation]. Equivalent with 1000 modern commercial airplanes: $10^{14} \times 10^3 = 10^{17}$ parts	

[a]The counts in electronic components in this table are from "Transistor count" (2016).

[b]The counts of cell components are from the BioNumbers Database (Milo, Jorgensen, Moran, Weber, and Springer, 2010, BNIDs 100071, 100267, 106198, 109387, 107552) and from Milo, Phillips, and Orme (2015, pgs. 4, 17, 25, 143).

plane. Similarly, the human body has 100 thousand times more parts than the aircraft carrier even though it weighs about 500,000 times less than the aircraft carrier.

3.3   Comparing the Level of Autonomy and Self-Sufficiency

Next we will look at machines in terms of their autonomy, using the definition below:

> **Autonomy** *is the characteristic of an entity to control itself and to operate independently of external factors.*

At the highest level, a car has no autonomy. The car just follows the commands of its driver in terms of direction, speed, acceleration, or breakage. The driver is in full control except when the car is in cruise mode. In this mode, the car maintains speed within the set speed limits, regardless of whether the car is running on flat or slopped terrain. Some may consider certain sub-systems of the car, such as the alternator, which charges the battery, or the compressor, which provides air conditioning, manifest a certain degree of autonomy. But this may be true only in regards to these systems starting and stopping their work function based on some regulating parameters such as voltage, air temperature, or humidity.

In general, genuine autonomy is present when an entity has many degrees of freedom for its operation and it has an internal decision-making capacity that determines which of the available options, or "directions," to take on each of the available dimensions of the decision space. Thus for a car, the human driver makes decisions on any of the available dimensions of car navigation: speed, acceleration, deceleration (breakage), and steering. As mentioned before, only while in cruise mode does the car have limited control of its speed.

A man-made machine that has (or promises to have) significant autonomy is a self-driving car, which can be seen occasionally on California freeways. After the human passenger keys in the destination, the self-driving car chooses its route to the destination and follows this route while successfully avoiding driving hazards including other cars running around, street lights, pedestrians, walls, and other obstacles. It is an instructive observation that the self-driving car exhibits this autonomy as a result of a complex set of supporting functions and, importantly, a unified control system that makes decisions based on various static and dynamic input data. Here are some of the supporting functions that are instrumental in the proper performance of the self-driving car:

- The motion function (acceleration and cruise control) provided by the car engine and transmission sub-machines

- The steering function provided by the steering sub-system

- The sensing function that senses the street, other cars, and objects[2]

The key sub-system of the self-driving car is the car control driving system whose role can be described as a combination of the following:

- Continuously collecting sensor and GPS positioning information

- Continuously comparing the car's driving status (e.g., speed, acceleration, direction) with the current position on the GPS route to the destination and subsequent correction of the driving parameters

- Continuously evaluating whether there are any driving hazards based on the sensing information flow and subsequent correction of the driving parameters to avoid collisions

What can be learned from the above analysis of the self-driving car is that genuine autonomy is very expensive to achieve from an engineering point of view because of the following:

- There are large numbers of degrees of freedom available during navigation including speed, direction, acceleration, route selection.

- There are large numbers of diverse types of constrains that need to be continuously observed including alignment to the GPS navigation path, continuous object proximity sensing, and collision avoidance.

- There are large numbers of supporting functions that must be precisely timed and controlled.

- There is a sophisticated control system that implements concurrent, sophisticated algorithms that evaluate continuous flows of information and make instantaneous decisions.

Let us now compare the autonomy of a self-driving car—a top achievement of human engineering—to the autonomy exhibited by a cat.

As a start, it makes sense to observe that the autonomy of the self-driving car maps to only one of the many aspects of the autonomy exhibited by a cat. This is the movement aspect or the ability of a cat to go to a destination starting from a certain, different location. And let us notice that in this comparison the cat's autonomous movement is analogous to *both* the passenger of the self-driving car who decides where the car's destination is and the car itself that autonomously reaches that destination. The cat is both the decision-maker and the achiever of the movement.

---

[2]This function is based on various types of sensors with which the car is equipped such as radar, video capture, distance evaluation, and GPS navigation. These sensors provide a continuous flow of sensory and positional information to the car's driving system.

Although there are differences between how a car moves and how a cat moves that do not allow for a perfect comparison between the autonomous movements of the two, it seems that the cat's overall degree of autonomous movement is much higher than that of a self-driving car. For one thing, the self-driving car's wheels must move on a relatively flat road that may have a limited slope, while the cat's movement path can go over fences, through ditches, valleys and crests, over the tree trunks or branches, on house roofs, or through subterranean narrow channels. The cat navigates a more varied terrain without significant challenges (including terrain it has never experienced), which means the cat's autonomous movement is superior to that of the self-driving car. This is because the cat can successfully "navigate" a greater number of dimensions of the movement and decision space. Another significant reason that the cat has superior autonomous movement is the much richer set of support functions and sub-systems that provide flexibility and versatility for cat movement, such as a rich set of environment sensing functions (e.g., visual, audio, touching) and body agility functions (e.g., walk, run, jump, climb and sneak).

To expand upon the comparison between the autonomy displayed by the self-driving car and that of the cat, the cat's autonomous movement—although superior to the car's autonomous movement—is an autonomous function that supports the cat's higher-level behavior and life-sustaining strategies (e.g., hunting for food, hiding for self-defense, fighting with rival cats, searching for food, or finding a mate). For all of these behaviors and strategies, the cat manifests an exceedingly superior level of autonomy. The cat is deciding when and how to look for food, when and how to look for a mate, etc.

The conclusion is that LOMs, and in particular the advanced organisms such as fish, birds, and mammals, exhibit an amazing level of autonomy that is far, far superior to that of even the best MMMs.

A related aspect to autonomy is that of self-sufficiency. An absolutely self-sufficient machine would be one that functions in perpetuity without any material support or condition provided to the machine from outside. Living organisms, although they have a remarkable level of autonomy and self-sufficiency, are still dependent on certain external materials and favorable conditions for their sustenance and survival. Basically, LOMs depend on the availability in their environment of the following materials and conditions:

- Air with life-supporting gases (e.g., O_2, CO_2, N_2) for terrestrial life or water with O_2 for aquatic life

- Food

- Alternating periods of sunlight and darkness

- Favorable environmental conditions such as favorable temperature ranges, transparent atmosphere, and protection from ultraviolet and cosmic radiation.

3.4   Comparing the Body's Structure and the Body's Nature

Man-Made Machines have mostly rigid components and hard bodies while Living Organism Machines have soft or at least pliable bodies.

MMMs might have certain components engaged in movements and kinematics relative to other MMM components, for example pistons, crankshaft, and valves in an internal combustion engine. In contrast, LOMs have soft bodies with both rigid and soft components. LOM bodies grow as the individual organisms grow. This is equally true for simple single-cell organisms and for multi-cellular complex organisms. Even the rigid components of LOMs, such as bones or shells, grow in harmony with overall growth of the organism with preservation of specific organism shape.

The softness of LOM bodies is a necessity because only a soft body with a variable volume and variable geometry allows for several important processes that are characteristic of LOMs:

- Self-replication of single-celled organisms and sexual reproduction of multi-celled organisms (Both processes require the development and growth of either a cloned cell or an embryo from the fecundated ovum to a mature organism.)

- Growth of the organism both prior to and after birth

- Replacement of cells in organs that have tissues requiring continuous cellular replacement

- Circulation and distribution of liquids (e.g., blood and lymphatic liquid) and gases to various organs and bodily systems

It is important to note that self-replication and organism growth are impossible if the organism has a fixed volume, fixed geometry, and rigid internal components. *The engineering problems that a growing, developing soft-bodied organism solves are significantly more complex than the engineering problems for designing and constructing a MMM with a rigid body and rigid parts.*

4   Unique Features of Living Organism Machines

Now that we have looked at the ways in which LOMs can be compared to MMMs, we will next look at the unique features that LOMs have that no MMM currently can implement.

4.1   Only Living Organism Machines Have the Ability to Self-Replicate

One defining characteristic of living organisms is their ability to self-replicate or reproduce. Organisms have many forms of reproduction. One of the simplest forms is the self-replication of single-celled organisms. The first phase of cell self-replication is where the cell grows by ingesting nutrients from the environment, and then a cloning process occurs where the cell's internal parts are replicated from the mother side of the cell to the daughter side. In the second phase, the cell divides, and the mother and daughter sides separate from each other. They become independent, fully-formed cells that are able to self-replicate again. Other forms of living organism replication, like sexual reproduction, can be much more complicated than the self-replication of single-celled organisms.

The Freitas Jr. and Merkle (2004) study of material self-replication identified three challenging conditions that must be met to implement physical Material Self-Replicators (MSR):

1. **Material closure.** This condition states that all materials—raw materials or refined/processed materials that are needed for the fabrication of all the self-replicator's components—must be available inside the Material Self-Replicator (MSR). More than that, the MSR should contain all of the processes that derive the required fabrication materials from the raw materials ingested by MSR from its environment.

2. **Energy closure.** This condition states that all energy in all forms that are needed inside the MSR to fabricate the components of the daughter MSR must be generated by the mother MSR from ingested materials.

3. **Information closure.** This condition states that all information needed to control the self-replication processes and all information that describes "what needs to be replicated and fabricated" must be fully available within the MSR.

In a previous study the requirements for designing the Simplest Self-Replicator (SSR) were outlined with the conclusion that implementing the simplest possible material self-replicator is a daunting engineering task that is (probably) significantly beyond the current, most-advanced engineering and scientific capabilities (Mignea, 2014b).[3] The study found that all three closure conditions, listed above, are extremely difficult to achieve. Other difficulties that we identified for implementing a SSR are:

1. The SSR must be fully automated so that all active components possess information processing and communication capabilities. This leads to the necessity that the SSR must be able to make a semiconductor fabrication facility on par with clean room fabs used by companies such as Intel or Qualcomm.

2. The SSR, as a computer, information, communication and software-driven system, will have a design on a level of complexity, reliability, and autonomy that

---

[3]See also the YouTube video presentation, "The Design of the Simplest Self-Replicator" from the *2012 Conference on Engineering and Metaphysics*, available at:
https://www.youtube.com/watch?v=dCqb_hyFHEA

is far beyond the most advanced human-built artifacts (e.g., a robotized car factory, a microprocessor manufacturing facility, etc.).

3. The design and the implementation of the SSR is particularly challenging because it must be a material object with variable volume and geometry so that the SSR can grow while the clone (daughter) SSR is increasing in size within the material enclosure.

4. The design and implementation of the SSR is also challenging because we do not possess nanoscale manufacturing tools that can manipulate the types of macromolecules that are the common building blocks inside living organisms.

4.2   Only Living Organism Machines Have the Ability to Self-Heal

The ability of many LOMs to repair themselves and heal after certain bodily injuries is absolutely remarkable. There is really no comparable capability in MMMs. Because it would take so much space to detail these amazing abilities, we will not go into detail on these capabilities. However, the literature on them is extensive.[4]

4.3   Humans as Living Organism Machines Have Exceptional Singular Abilities

Humans—the *Homo sapiens* species—have exceptional, singular capabilities that radically distinguish them from all other LOMs. Seen as machines, humans are super, super machines that exhibit "built-in" extraordinary capabilities that are so advanced that human engineers, scientists, and technologists cannot imitate them. Here is a summary of these extraordinary capabilities:

- Humans communicate through language.

- Humans are intelligent agents capable of reasoning and abstract thinking.

- Humans are capable of resolving an unlimited number of different types of problems.

- Humans are capable of imagining, designing, and constructing machines (MMMs).

---

[4]For starters, an interested reader can take a look at Mallefet and Dweck (2008), Alvarado (2000), and Michalopoulos and DeFrances (1997).

5   Scientific Miracles

5.1   The Challenge Presented by Living Organisms

When compared with man-made artifacts, living organisms can legitimately be seen as super machines that exhibit a rich set of extremely complex functions. LOMs have exceptional, singular abilities such as the ability to self-replicate or reproduce and the ability to self-heal and self-repair after illness or certain bodily injuries. LOMs are autonomous entities that exhibit numerous sophisticated self-supporting behaviors that roboticists and artificial intelligence engineers can only imagine. The remarkable practical intelligence of LOMs is based, among other things, on their exceptional ability to sense and perceive the environment (i.e., attend to its important aspects) and their singular motor skills that translate into efficient physical mobility and agility. The LOMs' nervous system and brain are essential elements that make them exceptionally well-integrated, dynamic, and efficient intelligent agents capable of developing life-sustaining and self-supporting behaviors.

We emphasize that many, if not most, of the LOMs' abilities are far, far beyond the abilities of man-made machines and artifacts that are made by the best scientists, engineers, and technologists.

All of these factors lead to the conclusion that living organisms are genuine scientific miracles. This conclusion is reinforced by a number of additional considerations.

Living organisms have been readily available for unmediated, direct examination, investigation, and internal scrutiny for centuries. However, in spite of significant progress in the life sciences (including perfecting investigative instruments as well as the huge accumulation of new biological knowledge), there are still a large number of unknowns regarding the internal composition, internal processes, and inner workings of living organisms.

There has always been an enormous interest in understanding the secrets of life. Interest in the vital mechanisms and information that animate living organisms is made manifest in the number of scientists and researchers and the vast amounts of material and financial resources that have been invested in the advancement of this knowledge. However, the paradox in the life sciences is that as more questions are answered, an even larger number of questions are raised.

There is an open question regarding the comprehensibility of more advanced living organisms or even the comprehensibility of their individual organs or their organs' systems. For example, is there an absolute or practical barrier to comprehending how the human brain works? It is estimated that the human brain, an organ with a volume of 1.4 liters and weighing approximately 3 pounds, has 100 billion neurons and 100-1000 trillion synapses (i.e., connections between the neurons). In spite of progress in understanding how electric potentials are propagated along neurons, synapses, and dendrites, there is a huge gap between understanding physical characteristics of measurable behaviors in the brain and knowing how capacities of the mind, such as memory, thoughts, and pattern recognition, emerge from the low-level brain elements and processes.

To illustrate one dimension of the challenges brain scientists face, consider the number of brain elements that need to be precisely observed and monitored in order to determine how the higher-level brain processes work. If we consider a synapse as equivalent to an elementary network switch, and estimate the Earth's population to be 10 billion people, then we estimate that there are 100 network switches (in wired or wireless equipment) for each person in the world. This gives a total estimate of the number of network switches in the world as 1000 billion switches, or 10^{12}. Now, considering the lower estimate of the number of synapses in the human brain as 10^{14}, then a single human brain has 100x more switches than all the network elementary switches in the world. The 100 multiplier computed this way may actually become a 100,000 multiplier if we consider the estimates of Stanford University professor of physiology Stephen Smith:

> One synapse, by itself, is more like a microprocessor—with both memory-storage and information-processing elements—than a mere on/off switch. In fact, one synapse may contain on the order of 1,000 molecular-scale switches. A single human brain has more switches than all the computers and routers and Internet connections on Earth.

(quoted in Moore, 2010)

Coming back to the issue of comprehensibility faced by neuroscientists, what is the approach for studying this organ that has 100 trillion synapses, or one hundred trillion microprocessor-equivalent units? In order to comprehend how the brain works, or at least to obtain a limited graph of connected neurons, synapses, and dendrites, even if you only needed to collect real-time data for only a segment of one millionth of the total brain, this still comprises about 100,000 neurons and 100 million synapses. But there are no such fine-grained neuronal brain probes today or coming in the near future. Even if such fine neuronal probes will exist, there are still overwhelming questions to answer:

- Will there be material space to insert such probes into the three-dimensional brain mass with precision and accuracy?

- How can the information from such probes be concurrently collected during the brain investigative experiment?

- How will the scientists know (before successfully running such investigative experiment) what is the brain sub-graph (connectome) whose crucial points (synapses) need to receive a micro probe for each point?

To emphasize the dimensions of the scientific challenge faced by neuroscientists, and to weigh if the human brain is a real scientific miracle, note that there are huge brain study projects both in the US and Europe that have been on-going for years. One example is the DARPA SyNAPSE Program, which can be summarized as:

> SyNAPSE is a *backronym* standing for Systems of Neuromorphic Adaptive Plastic Scalable Electronics. It started in 2008 and as of January 2013 has received $102.6 million in funding. It is scheduled to run until around 2016. The project is primarily contracted to IBM and HRL who in turn subcontract parts of the research to various US universities.

(Pearn, 2013)

The SyNAPSE program made significant progress toward some of its goals. One of the most important achievements is the TrueNorth chip, which is described as

> an IBM chip (introduced mid 2014) [which] boasts of 1 million neurons and 256 million synapses (computing sense); 5.4 billion transistors and 4,096 neurosynaptic cores (hardware).

(SyNAPSE, 2016)

Another example is The Blue Brain Project. In 2013 that project became the Human Brain Project (HBP), a billion-euro, ten-year initiative supported in part by the European Commission:

> The goal of the Blue Brain Project is to build biologically detailed digital reconstructions and simulations of the rodent, and ultimately the human brain. The supercomputer-based reconstructions and simulations built by the project offer a radically new approach for understanding the multilevel structure and function of the brain.

(In brief, 2015)

The Human Brain Project ran into problems two years after it started after certain scientists objected to its direction, after which the program changed its leadership, direction, and focus (Theil, 2015).

In spite of accumulation of significant new knowledge and developing a valuable, advanced system platform—both hardware and software—to test various hypothesis on how the actual human brain works, there is still skepticism that these studies, simulations, and experimental workbenches provide significant help in bridging the knowledge gap between the brain's material biology and physiology and how mental processes work at the abstract and conceptual level. In other words, it is not clear how the biology and physiology of the brain translates into thoughts, human intelligence, and consciousness. The challenge of understanding what consciousness is and how it emerges from the mind and brain is reasonably known as the **hard problem of consciousness** (Chalmer, 1995).[5]

---

[5]For a longer discussion on the many issues surrounding consciousness, see Gulick (2016).

5.2   A Look at Our Earth and Solar System

Before we consider the origins of the Earth and of the Solar System we propose considering them as cosmic scale, integrated machines. The motivation for considering these systems as machines is derived from the empirical rule that we stated before: *wherever there is sustained function, there is a machine.*

What are the sustained functions that the combination of the Earth, the Sun, the Moon and the other planets in the Solar System produce? Here are the important ones:

- Provision of light with regular succession of day and night and seasonal variations

- Provision of heat and energy

- Provision and retention of water reserves

- Provision and retention of air reserves and gases critical for life: O_2, CO_2, N_2

- Water transportation, water distribution, water recirculation, mineral transport by water

- Atmospheric and weather dynamics (preventing stagnation among other things)

- Protection of life from ultraviolet and cosmic radiation and other cosmic dangers

When we analyzed the Living Organism Machines we saw that in spite of their unequaled autonomy and high degree of self-sufficiency, they still have critical dependencies on certain substances such as air (O_2, CO_2, N_2), water, and favorable environmental conditions that include temperature, light, and a transparent atmosphere. It is interesting—and possibly just a coincidence—that all of the critical materials for life, as well as all of the critical features for favorable environmental conditions that are essential for LOM survival are provided by the Earth and Solar System. Even if this is a happy coincidence, the objective, undisputed reality is that the Earth and Solar System is the sustained provider of materials and conditions essential for life, and thus it is legitimate to consider the Earth and the Solar System as integrated super machines. *Why super machines?* Because of several reasons:

- Their huge, planetary scale

- The precision and reliability of the supplied functions lasting for ages

- The perfect match—coincidental or not—between the materials and conditions they provide and the critical needs of the Living Organisms Machines that they satisfy

Because of the perfect matching between the LOM super machines and the Earth and the Solar System super machine, it is reasonable to conclude that the web of LOMs inter-dependent ecologies in combination with the Earth and the Solar System together form what can be called the **Planetary Perpetuum Mobile for Life**. Indeed the planetary scale Earth and Solar System super machine, integrated with myriad types and specimens of LOM super machines, propagated and sustained life on Earth for ages in an autonomous, independent manner without any apparent help from any controlling or regulating agent.

5.3   Conclusion: Our Objects of Interest Are Genuine Scientific Miracles

We proposed the view that living organisms are super machines (and humans are super, super machines) and that life and living organisms are genuine scientific miracles. For all practical purposes LOMs are sophisticated super machines that are beyond current creative capabilities of humankind. The scientific institutions and research labs have plenty of matters to investigate, probe, and understand without any fear that the subjects of their interest will be exhausted at any foreseeable point.

We presented brief arguments that the Earth and the Solar System can legitimately be seen as super machines as well. It is not a stretch to view also Earth and the Solar System as scientific miracles because they, along with their myriad forms of life, form the Planetary Perpetuum Mobile for Life.

6   Evaluating the Origins Issue

6.1   The Mount of Complexity: A Background Metaphor for Evaluating the Origins Issue

The Mount of Complexity illustration presented in the drawing in Figure 8.3 is a graphical capture of the conclusions reached so far: the far superiority of the Living Organisms Machines (LOMs) versus Man-Made Machines (MMMs) that established one of the supporting arguments for characterizing LOMs as genuine scientific miracles.

The drawing shows on the left side an arrow pointing upward that represents levels of complexity, organization, functionality, and information, while at the same time shows an increase in scientific interest that the entities named above the markers represents.

At the base of the mountain are Man-Made Machines (MMMs) that move higher and higher up the mountain with increasing complexity and functionality. Thus at the bottom of the MMMs "altitude range" are the simplest machines (i.e., human-made artifacts) such as a wheelbarrow or a sailboat. At the top of the MMMs "altitude

Figure 8.3: The Mount of Complexity and Searching for a Way Matter Can Climb It

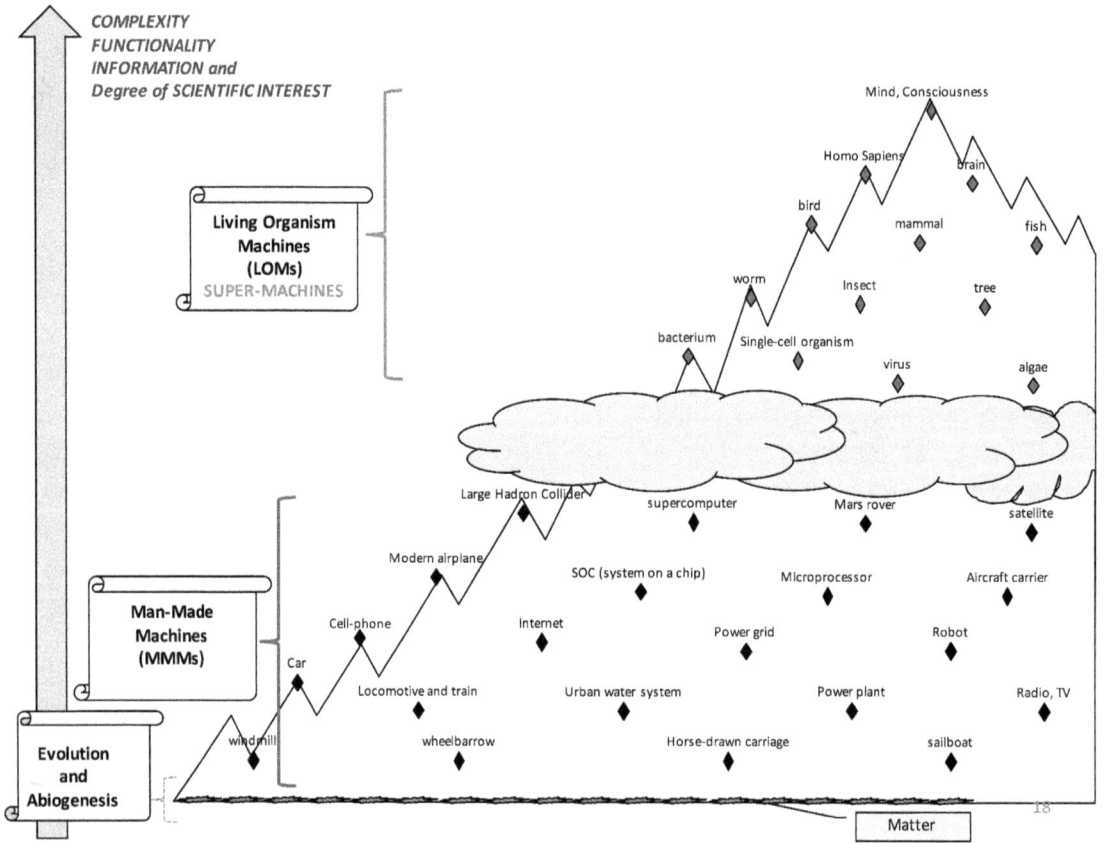

COMPLEXITY
FUNCTIONALITY
INFORMATION and
Degree of SCIENTIFIC INTEREST

Living Organism Machines (LOMs) SUPER-MACHINES

Man-Made Machines (MMMs)

Evolution and Abiogenesis

Mind, Consciousness
Homo Sapiens
brain
bird
mammal
fish
worm
Insect
tree
bacterium
Single-cell organism
virus
algae

Large Hadron Collider
supercomputer
Mars rover
satellite
Modern airplane
SOC (system on a chip)
Microprocessor
Aircraft carrier
Cell-phone
Internet
Power grid
Robot
Car
Locomotive and train
Urban water system
Power plant
Radio, TV
windmill
wheelbarrow
Horse-drawn carriage
sailboat
Power plant

Matter

range" are some of the most sophisticated human-made artifacts such as the Large Hadron Super Collider, the supercomputer, or the Mars Rover.

At a much higher altitude on the Mount of Complexity is the altitude range for Living Organism Machines (LOMs). It is separated by the MMMs altitude range by clouds that are meant to suggest the huge gap in complexity between MMMs and LOMs. There are markers of the various living organisms of varying complexity. For example, at the bottom of the LOMs' altitude range, we find viruses and single-celled organisms, while at the top of the range we see the signposts for the brain, *Homo sapiens* species, the mind, and consciousness.

At the absolute bottom of the Mount of Complexity, there is a graphic that represents matter—unstructured and uninteresting "stuff"—and the starting point for any ascent up the Mount of Complexity. Materialism and methodological naturalism must provide us with convincing theories of origins that persuade us that matter is able to climb the Mount of Complexity and create the entities in the LOMs' altitude range. *In other words, materialism must show us how matter, and whatever instruments of change are in its creative toolbox, managed to create machines that are not only far, far superior to the most sophisticated human artifacts, but are also genuine scientific miracles that have mystified armies of scientists and researchers for centuries.*

The Mount of Complexity serves as a graphical illustration of the background metaphor for evaluating origins narratives and theories.

6.2   We are surrounded by scientific miracles

The materialist and the methodological naturalist are faced with an undeniable, objective reality: living organisms, life, the brain, the mind, consciousness, and the Perpetuum Mobile for Life hosted by Earth and our Solar System are genuine scientific miracles. They are transcendental because they are clearly beyond the creative abilities of human beings. They are supernatural because we cannot identify any entity in our natural world, including the human mind, that is capable of creating such entities.

6.3   The Materialist Proposition for the Origin of Life and of Living Organisms

We know from our experience that only humans possess the intelligence required to imagine, design, and build machines. There are animals and birds that have their own kind of intelligence that allows them to build shelters, nests, and sometimes use objects as ad hoc tools to solve life sustaining problems. [6] However, living organisms are super machines that have functions, internal architectures, and the ability to build that are beyond the abilities of human engineers.

We turn now to how methodological naturalism and materialism's narratives explain how living organisms, and life in general, originated and became what they are today in their countless forms.

6.4   The Instruments of Change in Materialism's 'Creation Toolbox'

Let us consider the logical foundations as well as the explicit and implicit assumptions that materialist and methodological naturalists use in articulating their creation story.

First there is the explicit strong assumption that matter is the supreme and the unique component whose properties explain all phenomena in our world including life and living organisms. MN excludes any consideration of a supernatural intelligence when conducting science and thus when trying to explain the origin of life and living organisms.

---

[6]For a good example, see "How smart is a crow," a *BBC* 4 minute clip on YouTube: https://www.youtube.com/watch?v=URZ_EciujrE

The biggest problem the materialist scientist faces is that it is very difficult, if not impossible, to ignore the objective complexity of life, including living organisms, the brain, the mind, and consciousness. It is strikingly evident that life and living organisms are the super machines that they are because their exquisite, rich functions cannot be realized without an underling design that includes organization, information, and the purposeful arrangement of parts and subsystems.

The materialist and methodological naturalist are left searching for any means, materials, or natural processes that do not assume intelligence or purpose and can somehow climb the bottom-up path from inert matter to structure, to organization, to arrangement of parts, to design, to machine, and to function.

So, where can the materialist look to find help for his search to ascend this path? They can only look to particles and phenomena in the inanimate world that are able to bring some kind of action, change, transformation, or some way of putting things together to make something of a different quality. This something of a different quality would have to be something that can be objectively placed a little higher up on the Mountain of Complexity.

A classic example of such an instrument of change is the lightning that Miller and Urey simulated with electrical sparks in their famous experiment to probe if some advanced organic chemical compounds can be generated spontaneously in nature (Miller and Urey, 1959).

We are inventorying here the elements that materialism invokes frequently as instruments of change in their research and narratives regarding origin of life (OOL). It is important to have an idea of what these instruments of change are and how these instruments acted to presumably lead to the appearance of the first life forms on Earth. The materialist instruments of change include:

- Environmental phenomena:[7]

    - Rain, snow, hail, freezing, melting, flooding, drought
    - Lightning, tornadoes, gusts of winds
    - Earthquakes, volcanic eruptions, mudslides, lava flows
    - Underwater thermal vents, underwater volcanic eruptions
    - Fires

- Chemical reactions:

    - combustion, neutralization, redox

- Physical events:

    - Collisions, objects rolling downhill, icebergs melting, water boiling

---

[7]There is an implicit assumption here that the same environmental phenomena that exist today were present when life was supposed to appear spontaneously on Earth.

- Sublimation, magnetization, irradiation

- Cosmic events:

  - Meteorites, comets hitting the Earth, cosmic radiation

- Liberal amounts of chance and luck, fortuitous circumstances

The materialist, who affirms that matter can explain everything, creates for himself an impossible handicap for his methodological naturalism-aligned science. There is no recorded event in known human history where the interplay of the instruments of change identified above were able to assemble anything resembling a machine, no matter how simple and primitive. But life and living organisms are incredibly complex and sophisticated machines.

Given the rigid constraints on science imposed by methodological naturalism, materialists have no other option but to declare that life is a manifestation of matter and to channel enormous scientific resources into finding plausible hypotheses on how life could have originated on Earth as some fortuitous sequence of natural phenomena and material events. There are enormous problems that stand in the way of any such abiogenesis scenario.

6.5   Logical Requirements of Material Self-Replication

One of the main requirements for a material entity to be considered alive is the ability to self-replicate. Following the self-replication pattern of the simplest, single-celled organism, self-replication follows a two phase process:

- A cloning phase during which the entity grows and creates inside itself a copy (the daughter) of the original (the mother) entity.

- A division phase when the daughter entity reaches "maturity," separates from the mother entity, and becomes an independent living organism that can self-replicate.

In Mignea (2014a) we studied material self-replication and arrived at a significant list of functional requirements for a material entity to be able to produce an exact replica of itself.[8] Here we provide a simplified version of the logical steps, along with their functions and components, that must be present in the Simplest material Self-Replicator (SSR).

1. The SSR must be able to grow in volume to follow a two-phase replication process, cloning and division.

---

[8]See also the YouTube video presentation, "The Design of the Simplest Self-Replicator" from the *2012 Conference on Engineering and Metaphysics*, available at: https://www.youtube.com/watch?v=dCqb_hyFHEA

2. The SSR must transform ingested matter into the components of the daughter SSR (i.e., the clone), which is the metabolism requirement.

3. The SSR must have an enclosure to isolate and protect itself from the environment.

4. The SSR must have an input gateway to accept and ingest materials for internal growth and replication.

5. The SSR must have an output gateway to eliminate the refuse material left after metabolism.

6. The SSR must produce energy from metabolizing input materials. This energy is needed to power all SSR internal processes.

7. The SSR must somehow create an exact replica of itself: a daughter SSR (or SSR clone) identical with the original SSR (mother SSR). This requirement has the following sub-requirements:

   a. The SSR must have the ability to fabricate all internal parts of the SSR including the fabricator.

   b. The SSR must have the capability to "know" how to copy itself.

   c. The SSR must have some powerful self-reflective, self-knowledge capability. This means, for example, that:

      i. SSR must know the list and nature of all its components, how are they are organized, and how they relate to each other. OR maybe

      ii. The SSR has a (mechanical) ability to systematically explore itself and identify the nature of all its components, accurately copy each component, and then engage the cloned components to the other cloned components so that they function in similar relationship to the original components.

   d. During cloning the SSR must know how to discern the cloned parts from the similar original parts.

   e. The SSR must know how to construct and correctly assemble all cloned parts into the emerging daughter SSR.

   f. The SSR must know how to sequence, schedule, and coordinate all actions: production (through metabolism) of primary materials, fabrication of parts, and assembly of parts into sub-assemblies and into the final daughter SSR.

   g. The SSR must know when and how to trigger the separation of the cloned daughter SSR from the mother SSR after the daughter SSR has been fully constructed.

h. The SSR must know how to copy all its "know-how" into the daughter SSR before such separation and then to "start the engine" of the daughter SSR as the last act of separation.

Our study of the simplest self-replicator (SSR), a material object to be designed and constructed by human engineers with ability of material self-replication, arrived at the following conclusions:

- A material self-replicator (SSR) is an enormously complex machine.

- The SSR is beyond current engineering technologies in many areas: system design, information design, hardware and software design, nanotechnology, material closure, energy closure, and information closure.

- A SSR cannot be constructed today by human engineers with the best available technologies and know-how.

6.6   Current Status of Origin Of Life (OOL) Research

There have been tremendous research efforts in terms of talent, material, and financial resources spent on OOL research. In spite of decades of such efforts, there has been no real progress in providing a credible and verifiable hypothesis on abiogenesis. Thousands of approaches and hypotheses have been formulated and tested, but one after another, they fail to provide real progress toward success (Thaxton, Bradley, and Olsen, 1984; Meyer, 2009).

This is not surprising that all of these efforts are not successful. In our view, there are fundamental reasons that explain current and future failures in the naturalistic origin of life research:

- There are no records that inanimate nature can create any non-trivial machine (i.e., more than three or four parts).

- The instruments of change in the materialism "creation toolbox" are too dull and dumb.

- Our experience, intuition, and reason tell us that only intelligent agents are capable of building intricate machines.

Most OOL research can be classified into three categories: replicator first, metabolism first, or membrane first. But none of these approaches is realistic since a functioning self-replicator must be at the same time a self-replicator, a metabolizer, and enclosed in a membrane. In other words, all three categories of components of the machine must be present and assembled in the first place.

7   The Materialist Creation Story

7.1   The Darwinian Theory of Evolution

The creation story employed by materialists to explain the enormous diversity of life is the Darwinian Theory of Evolution (TOE). Here are the main precepts of the ToE:

- Life in all its myriad of forms developed from one or a few primitive forms of life.

- The mechanism for the ToE is a combination of errors or accidents (i.e., random mutations) with natural selection.

- The ToE proposes that this mechanism generated more advanced organisms from simpler ones by improving them, making more complex species, and developing new body plans.

It is important to carefully consider the ToE's claims. Let us consider the hypothetical Most Primitive Life Form (MPLF). Our comparative study of LOMs and of the amazing complexity of the simplest self-replicator led us to conclude that even the MPLF must have been an enormously complex machine that functioned with extreme precision since it had the ability to self-replicate along with all of the other complex functions of a self-supporting life form.

The ToE claims that a sequence of errors or accidents (random mutations) in the internals of the MPLF led to an Improved Life Form. This is logically analogous to the claim that fabrication errors in the gear box of a car can lead to an improved car. Or it is like bending an arbitrary tooth of a gear in a clock and having it result in a more advanced or precise timepiece. Or it is logically analogous to the claim that some random scrambling of the lines of code in a computer program will make the program better. Fundamentally, stating that random changes in a well-functioning machine can have any other outcome than that the machine stops functioning, or at a minimum that the machine has a degraded manner of functioning, is pure folly and shows a disconnect from reason and day-to-day experience.[9]

The second active element in the ToE is Natural Selection (NS). The most prolific machines that were subjected to accidental changes (i.e., random mutations)—those that left the most numerous offspring—conform to NS and therefore become the most abundant varieties of machines within that species.

Natural selection makes sense in one way. Among all machines of the same class (i.e., among all organisms of the same species), the machines that are most

---

[9]A small exception to this is shown by Behe (2007) where, in what Behe calls "trench warfare," organisms can *discard* functionality to improve local outcomes. However, this is far from being able to explain the *origin* of functional machines to begin with.

likely to leave the most numerous descendants are those machines that suffered the smallest number of accidental changes (i.e., the fewest errors). In other words, NS—if it ever has significant effects—is eliminating the machines (the organisms) that had the largest number of errors (random mutations) and preserving those that were least subject to these errors. But this is in stark contrast with the ToE, which requires a large number of random errors to produce more and more advanced organisms. Then the two main mechanisms of the ToE, random mutation (RM) and natural selection (NS), are logically and fundamentally antagonistic to each other, and thus the ToE cannot be effective in leading to more advanced organisms.

There are numerous other and substantial critical arguments against the ToE. However, the principal and fundamental argument against ToE is that errors cannot produce anything other than degradation and failure in a functioning machine. There is no need to consider many other substantial arguments against the ToE. The ToE fails irreparably from the start with a plain examination of "random mutations" as a creative mechanism for ToE. What is surprising is that ToE has been accepted as a scientific theory even though it is founded on such an irrational claim as errors having mysterious and supernaturally creative capabilities.

In summary the ToE is built upon two flawed principles that are clearly incapable of producing the amazing diversity and exceeding complexity of living organisms.

To conclude, our analysis of the primary mechanisms of ToE recognizes that RM and NS express a fundamentally absurd and illogical belief that deserves its own name: the *ToE paradox of the creative power of errors*. In essence the two principles, RM and NS of the TOE mechanism for evolution, can be formulated in this way: *the larger the number of errors inflicted upon an organism, the more advanced and unique is the resulting organism (if they pass through a selection process).*

7.2 The Materialism Creation Story for the Earth and the Solar System

Materialists have no other choice, given the foundation claim of their dogma *that everything can be explained as manifestations or results of matter*, than to look to matter to resolve all scientific conundrums. This is likely why we see the evolution narrative used not only in the realm of biology but also in other diverse domains such as cosmogony, sociology, psychology or economics.

We are going to look briefly at the creation story for our Universe, Solar System, Earth, Sun, and the planets proposed by materialism and methodological naturalism. While random mutations may have made some sense in biology, it is not clear what the mechanisms for transformation for a supposedly continuous 'evolution' towards a more advanced organization, structure, and complexity of the Universe or our Solar System are.

The mainstream scientific theory of cosmic-scale origins starts with the Big

Bang 13.7 billion years ago, and is followed by the formation of the bodies within the Universe from gas and dust clouds generating stars, galaxies, and other solar systems. Stellar fusion processes generated chemical elements with greater and greater atomic masses, such as iron from hydrogen and helium, which eventually lead to the formation of solar systems and planets like Earth where heavy elements dominate their composition. Then, somehow, water accumulated on Earth and combined with the Sun's heat and light, it was inevitable that the most primitive forms of life would appear on Earth. From there, it was the job of evolution to complete the story by creating the diversity of life that we see around us today.

The materialist creation story of the Universe, our Solar System, and Earth does not provide an answer to the following conundrum: the current state of the Universe has significantly more order, structure, and information compared to the moment the Big Bang happened. Materialist cosmology cannot explain how this injection of order, structure, and information happened while also maintaining the second law of thermodynamics, which states that the current level of entropy (the amount of disorder) should be much, much greater than the entropy at the moment of the Big Bang (Penrose, 2006).

In plain language the question is how can an explosion of cosmic proportions—we can think of Big Bang to be such a thing—give birth to structured galaxies, stars, and planetary systems? In particular, how did our Milky Way Galaxy originate? How did our Solar System and Earth originate? Isn't there a miraculous coincidence that life and Living Organism Machines appeared on a rocky planet with the proper amount of surface water, with a liquid magnetic core that provides life protection for ultraviolet light and cosmic radiation but at the same time has a proper atmosphere and revolves around the proper type of star at a proper distance for favorable temperatures? Isn't it a remarkable coincidence that the dependencies of living organisms in terms of air, water, proper environment conditions, the succession of light and dark periods is perfectly provided by the Solar System and the Earth? Isn't it reasonable to observe that the myriad of living organism machines' dependencies are perfectly satisfied within and among themselves, but also by the Earth, Sun, Moon and our Solar System that makeup a complementary cosmic-scale machine?

It is interesting to observe that the mainstream scientific theories (such as accretion theory) on star formation, galaxy formation, and the formation of our Solar System (including our Earth and Moon formation) have many unsolved problems (Finkbeiner, 2014). The cosmic "instruments of change" such as gravitation, the Big Bang, inflation, dark matter, and dark energy cannot provide a clean account of the origin of our Solar System and the Earth. In the same way that materialism is incapable of creating the Living Organism Machines, the materialist's cosmic instruments of change are too dumb to provide a credible explanation for how this cosmic-scale machine originated in a perfect form to be an essential part of the Perpetuum Mobile for Life.

8 Conclusion

There have been amazing advances in science and technology in the last century and especially in the last decades. There has been significant, extended, and in-depth knowledge accumulated about the objects of interest (e.g., life and living organisms, the Earth, our Solar system, the Milky Way Galaxy). Studies on human anatomy, physiology, health, and life extension developed into an ongoing paradox: unlike other phenomena, as we accumulate massive amounts of new knowledge, the sphere of unknowns becomes larger. This sphere of unknowns is the unanticipated details, subtleties, processes, and mechanisms that govern human life and biological processes. For example, the efforts made by many fields of science to understand how the human brain works have not brought us any nearer to achieving a holistic understanding in this domain. The hard problem of consciousness becomes even harder when even understanding how the brain—and not even necessarily the mind—works has become a farther removed possibility. What is the answer to fundamental questions like where does the information for the body plan of an organism reside? What are the developmental biology secrets that govern the exquisite complex processes of embryonic and fetal development?

The perplexing observation from the life sciences can be properly summarized in a short but rich affirmation that is hard to contest: *life and living organisms are genuine scientific miracles*. With equal justification, organs, organ systems, and the human brain in particular, can be considered scientific miracles.

Undoubtedly the idea of scientific miracles will be resisted by the materialist who takes to heart the core dogma of methodological naturalism. However our proposal of conferring the quality of scientific miracles to living organisms is the logical conclusion of an investigation conducted in a materialist-friendly manner. Is there anything more familiar to the materialist than the concept of a machine? Any machine—including Living Organism Machines—is made of material parts and components with real physical interactions that produce sustained functions, which can be objectively observed. Our empirical corollary that *behind any sustained function there is a machine producing it* also cannot be denied by the materialist. Otherwise the materialist must provide a replacement non-mechanical explanation or justification for how that observable function is produced.

Because of the many scientific miracles documented here, we suggest that there is an existential challenge to the scientific comprehensibility of the world around us, in particular of the living world, living organisms, and especially the human brain, the human mind, and consciousness. Another way to formulate this idea is the possibility that certain things may be practically *incomprehensible* for science. The speculative idea of incomprehensibility may be connected with the concept of *inscrutability*, i.e., the inability to shed light on certain life phenomena and life processes and, thus, leave things in obscurity. In many of these fields, additional data is not leading any closer to an answer.

The potential problem of scientific incomprehensibility, or inscrutability, is that there might be processes and phenomena (of a physical-chemical or of a different nature) in living organisms at the molecular, cellular, organ or whole organism level that are currently not known to science but are fundamental for the function of the living organism. Or such phenomena or processes happen at such a scale that scientific investigation becomes very difficult if not practically impossible (e.g., quantum processes).

Other aspects in science and technology that are relevant for our study is the belief of certain scientists and engineers that efforts in artificial intelligence (AI) will never be able to mimic the human mind. Thus far, there have been several historical disappointments in the field of artificial intelligence. Although there are AI advocates who believe in a golden future for AI, there are serious scientists and engineers that remain skeptical that AI will ever be able to approach the capabilities of natural human intelligence. As we mentioned in the text the native perception and functional capabilities of a cat, for example, are far, far superior to the capabilities of the most advanced robots. This way at looking at technology reinforces the thesis that we are surrounded by genuine scientific miracles.

In the text, we observed that, interestingly, the few requirement of autonomous living organisms in terms of air, food, water, and a favorable, life-supporting environmental are—coincidentally or not—supplied by the local cosmic machinery comprised of the Earth, the Sun, the Moon, and the rest of our Solar System. We also talked about looking at how Living Organism Machines (LOMs) and the planetary machines in the Solar System are highly integrated and co-dependent, constituting a **Planetary Perpetuum Mobile for Life**. This appears to be an autonomous system that has supported varieties of life for ages, and it is from these observations that we suggest that this Planetary Perpetuum Mobile for Life can be legitimately seen as another unique scientific miracle.

We contrasted the unimaginable complexity of living organisms with the lamentable origins' theories for the creation of these entities proposed by materialists and methodological naturalists.

Materialism has no other option but to explain origins from the bottom-up, and to search for and imagine any means by which matter may climb the colossal Mount of Complexity. Materialism cannot enlist intelligence into its ascent up the mountain. Consequently, the materialism story of creation cannot invoke anything that is associated with intelligent agents including goals, purposes, plans, schemes, or designs. Materialism is left to invoke the impotent instruments of change in its creative toolbox, such as lightning, earthquakes, or meteorites. Materialism may also search for or invent hypothetical, exotic mechanisms or processes that have the ability to create order, structure, information, machinery, and function from chaos, luck, and matter.

Materialism and methodological naturalism painted themselves into an unenviable, uncomfortable corner. They claimed that they can start from the bottom—from where raw matter exists—and figure out a credible way to climb the overwhelming Mount of Complexity to its highest peaks, which feature life, living organisms, mind, and consciousness. Materialism registers as a total failure, a resounding defeat, commensurate in size only with the enormity of its claim:

> ...physical matter is the only or fundamental reality and that all beings and processes and phenomena can be explained as manifestations or results of matter.

(materialism, 2016)

The paradox of materialism is clear: it eliminated the supernatural as a starting scientific hypothesis, but it is overwhelmed by the supernatural features found in scientific conclusions.

Materialism proclaims that matter is supreme and fundamentally removes intelligence from consideration. Intelligence is the single known phenomenon that can create structure, organization, information, machinery, and function.

Materialism creates its own mythology where random events collaborate with the goddess Chance to create the first life forms. The mythology consistently employs upside-down logic where errors, mistakes, and damage do not lead to degradation, destruction, and death. Instead, they have mysterious, superlative creative powers that take the most primitive forms of life and transform them—like a frog transforming into the Prince Charming—into wonderful, diverse organisms that overwhelm scientists with their unbounded complexity.

The materialists and methodological naturalist strongly reject any supernatural intelligence as a starting hypothesis for scientific inquiry. But it is becoming harder and harder, to the point of becoming practically impossible, for them to deny the transcendental and supernatural signature that is heavily imprinted on life, living organisms, the brain, the mind, consciousness, and the planetary scale Perpetuum Mobile for Life. All of these are some of the most interesting topics of scientific inquiry.

Materialism replaces the creative powers attributed in traditional cultures to a supernatural, intelligent Deity with the mythological, mysterious supernatural creative powers of matter.

References

Alvarado, A.S. 2000. Regeneration in the metazoans: why does it happen? *BioEssays* 22(6):578–590.

Bartlett, J. 2014. Using Turing oracles in cognitive models of problem-solving. In J. Bartlett, D. Halsmer, and M.R. Hall (editors), *Engineering and the Ultimate: An Interdisciplinary Investigation of Order and Design in Nature and Craft*, pp. 99–122, Blyth Institute Press, Broken Arrow, OK.

Behe, M. 2007. *The Edge of Evolution: The Search for the Limits of Darwinism*. Free Press.

Chalmer, D.J. 1995. Facing up to the problem of consciousness. *Journal of Consciousness Studies* 2(3):200–219.
http://consc.net/papers/facing.html

Finkbeiner, A. 2014. Astronomy: Planets in chaos. *Nature* 511:22–24.
http://www.nature.com/news/astronomy-planets-in-chaos-1.15480

Fodor, J. 2001. *The Mind Doesn't Work That Way*. Bradford Books.

Forrest, B. 2000. Methodological naturalism and philosophical naturalism: Clarifying the connection. *Philo* 3(2):7–29.

Freitas Jr., R.A. and Merkle, R.C. 2004. *Kinematic Self-Replicating Machines*. Landes Bioscience.
http://www.MolecularAssembler.com/KSRM.htm/5.6.htm

Gulick, R.V. 2016. Consciousness. In E.N. Zalta (editor), *The Stanford Encyclopedia of Philosophy*, winter 2016 edition.
http://plato.stanford.edu/entries/consciousness/

In brief. 2015. *The Blue Brain Project* .
http://bluebrain.epfl.ch/page-56882-en.html

Mallefet, P. and Dweck, A.C. 2008. Mechanisms involved in wound healing. *The Biomedical Scientist* pp. 609–615.
http://www.dweckdata.co.uk/Published_papers/Wound_Healing.pdf

materialism. 2016. *Merriam-Webster Online Dictionary* .
http://www.merriam-webster.com/dictionary/materialism

methodological naturalism. 2016. *Conservapedia* Revision 1258260.
http://www.conservapedia.com/Methodological_naturalism

Meyer, S.C. 2009. *Signature in the Cell: DNA and the Evidence for Intelligent Design.* HarperOne.

Michalopoulos, G.K. and DeFrances, M.C. 1997. Liver regeneration. *Science* 276:60–66.

Mignea, A. 2014a. Developing insights into the design of the simplest self-replicator and its complexity: Part 1—developing a functional model for the simplest self-replicator. In J. Bartlett, D. Halsmer, and M.R. Hall (editors), *Engineering and the Ultimate: An Interdisciplinary Investigation of Order and Design in Nature and Craft*, pp. 169–186, Blyth Institute Press, Broken Arrow, OK.

Mignea, A. 2014b. Developing insights into the design of the simplest self-replicator and its complexity: Part 2—evaluating the complexity of a concrete implementation of an artificial ssr. In J. Bartlett, D. Halsmer, and M.R. Hall (editors), *Engineering and the Ultimate: An Interdisciplinary Investigation of Order and Design in Nature and Craft*, pp. 187–212, Blyth Institute Press, Broken Arrow, OK.

Miller, S.L. and Urey, H.C. 1959. Organic compound synthesis on the primitive earth. *Science* 130(3370):245–251.

Milo, R., Jorgensen, P., Moran, U., Weber, G., and Springer, M. 2010. Bionumbers—the database of key numbers in molecular and cell biology. *Nucleic Acids Research* 38(supplement 1).
http://bionumbers.hms.harvard.edu/

Milo, R., Phillips, R., and Orme, N. 2015. *Biology by the Numbers.* Garland Science.

Moore, E.A. 2010. Human brain has more switches than all computers on earth. *Cnet News* .

Pearn, J. 2013. Darpa synapse program. *Artificial Brains: The quest to build sentient machines* .
http://www.artificialbrains.com/darpa-synapse-program

Penrose, R. 2006. Before the Big Bang: An outrageous new perspective and its implications for particle physics. In *Proceedings of EPAC*, pp. 2759–2762.

SyNAPSE. 2016. *Wikipedia* Revision 733393465.
https://en.wikipedia.org/wiki/SyNAPSE

Thaxton, C.B., Bradley, W.L., and Olsen, R.L. 1984. *The Mystery of Life's Origin: Reassessing Current Theories.* Lewis and Stanley.

Theil, S. 2015. Why the human brain project went wrong—and how to fix it. *Scientific American* .
http://www.scientificamerican.com/article/why-the-human-brain-project-went-wrong-and-how-to-fix-it/

Thiel, P. and Masters, B. 2014. *Zero to One: Notes on Startups, or How to Build the Future.* Crown Business.

transistor count. 2016. *Wikipedia* Revision 754563608.
https://en.wikipedia.org/wiki/Transistor_count

9 ‖ Problems With Non-Naturalistic Theories of Science

Eric Holloway

Baylor University

Abstract

Non-naturalistic theories of science are often criticized. The critics usually do not criticize the theory itself, but either misrepresent it or criticize the implications they consider wrong. What these theories need is internal criticism. A well-known and well-defined non-naturalistic theory is Intelligent Design (ID). There are currently no criticisms of ID theory from ID proponents. However, there are numerous legitimate issues with the theory such as difficulty in testing the theory and the theory requires fundamental revisions in our view of reality. It is important that these problems are explored in order to improve the usefulness of the theory. The criticisms will focus on ID, but similar criticisms apply to all non-naturalistic theories.

1 Categories of Criticisms of Non-Naturalistic Theories

There have been numerous criticisms of non-naturalistic theories. Examples of such theories are libertarian free will, vitalism, formal and final causality, and Platonic mathematics. These theories all have the common characteristic that they assume there is more to existence than the material world. However, in modern science, the goal is often to explain everything by material causes.

This trend is known as "methodological naturalism." Methodological naturalism states that only natural causes can explain natural effects. In this paper, "natural" is defined to mean anything that can be described by the laws of physics. Since the laws of physics can be described by a universal Turing machine, this means the natural world is describable by a universal Turing machine, also referred to as an algorithm. The set of entities describable by an algorithm is a super set of the natural world, so a general result that applies to algorithms also applies to the natural world and, consequently, methodological naturalism.

Additionally, sometimes the term "materialism" will be used in this paper. For the purposes of this paper, it means the same thing as naturalism.

The criticisms of ID come in three main categories: the implications of ID, ID's technical issues, the logic of ID.

1.1   Criticisms about the Implications of Intelligent Design

The first category focuses on arguments that criticize the implications of ID. One example is the neo-Thomist criticism that ID imposes a mechanistic view of reality (Beckwith, 2010). Another is that ID is young-Earth creationism in disguise since it implies the existence of God (Melott, 2002). These criticisms, if correct, do not imply that ID is a false theory.

1.2   Criticisms Dealing with Technical Issues

The second category of criticisms deals with technical issues. These criticisms detail incorrect or unclear theory and its applications. David Wolpert states that Dembski does not fill in all the necessary details to apply the No Free Lunch Theorem (Wolpert, 2003). Others state that the application of ID to evolutionary algorithms is irrelevant for biological evolution because the target for evolution is not fixed (Chu-Carroll, 2006). These criticisms, like those in the first category, do not prove ID false if they are correct. Neither do these criticisms mean ID is unfalsifiable. However, they may mean ID is not clearly enough articulated to be falsified.

1.3   Criticisms Addressing the Logic of Intelligent Design

The third category of criticisms addresses the logic of ID. If successful, these criticisms are capable of proving ID wrong. Dawkins makes this kind of criticism when he claims ID requires ever increasing unexplained complexity. As a result ID does not explain anything. He states that the creator must be even more complex than the creation, thus requiring even more explanation (Dawkins, 2008).

Elsberry and Shallit state that the explanatory filter identifies many objects with known naturalistic or algorithmic origins as intelligently designed (Elsberry and Shallit, 2011). They give examples such as stones smoothed by the sea that look like pottery and mountain crags that look like face silhouettes. They developed algorithms that produce bitstrings to exhibit complex specificity and an original metric called

Specified Anti-Information. These criticisms are more significant since, if successful, then ID is a false theory.

However, these criticisms tend to be marred by misapplication and misunderstanding of the theory. ID is meant to explain the origination of complex specified information (CSI). The trend of CSI from creator to creation must consequently increase. However, if the creator is more complex than the creation, this causes the amount of CSI to decrease going from creator to creation. Furthermore, if artifacts are regularly created by natural or algorithmic processes, then they are explained by the chance hypothesis and/or the specification is not separable from the process. If either are true, then the artifacts do not possess CSI.

There are numerous valid criticisms in the first and second category. These criticisms point out how at odds ID is with many commonly held theories as well as the difficulties in applying ID. The following criticisms will be identified with the category they fall in, either **1** or **2**.

To this author's knowledge, there are no coherent criticisms of the third kind that outright disprove ID. However, there is a fourth category that is similar to the third category. It deals with shortcomings in the theory as it currently is, but these shortcomings do not disprove the theory. These criticisms will be identified with **3a**. The criticisms will not be grouped by their category, as the criticisms build on each other out of category order.

2   The Criticisms

In the following criticisms, all that deal with non-naturalistic theories apply to ID. However, not all criticisms that apply to ID will apply to other non-naturalistic theories.

2.1   Unembodied Cause (category 3a)

Criticism: *Non-naturalistic theories need causes that regularly interact with the physical world in order to be detected.*

All sciences are empirical sciences. They work by proposing causes for effects, and then go out to detect these causes. If the cause is unembodied, then it is difficult to make it the subject of empirical science. Even immaterial phenomena, such as magnetic fields and electromagnetic waves, are embodied in the sense that their presence can be detected from their impact on matter. Non-naturalistic philosophies of science need to start with embodied causes.

How can a cause be both embodied and non-naturalistic? This sounds like a contradiction in terms. However, naturalism is not identical to embodiment. For instance, metaphysical naturalism is used to rule out the existence of souls. Souls are presumably embodied. So if souls exist, then metaphysical naturalism is false. This shows that something can be embodied while it is not naturalistic.

More specifically, naturalism as defined by the laws of physics can be violated by embodied agents. In general, this takes the form of violating various conservation laws. If an embodied agent violates a conservation law, then the agent is non-naturalistic.

In the area of ID theory, the conservation law of concern is the conservation of information. If an embodied agent can violate this law, then the embodied agent is a non-naturalistic intelligent agent.

2.2 Active Information Creation (category 2)

Criticism: *Non-naturalism needs to describe a causal agent that can create active information.*

Active information is a concept from ID theory, and it applies to search and optimization algorithms (Dembski and Marks II, 2010). *Active information* is defined as the information that allows a search algorithm to find a target more efficiently than random sampling. The principle of *conservation of information* states that active information cannot be created by algorithms. If methodological naturalism is true, then everything in the physical world can be described by an algorithm. Therefore, if methodological naturalism is true, active information cannot exist.

By *modus ponens*, if active information does exist, then methodological naturalism is false. The problem is that we do not have a good description of a cause that can create active information.

Despite this descriptive problem, we do know that active information exists. A specific instance of active information is the field of machine learning. Machine learning consists of choosing a hypothesis using labeled data that can predict the labels on new data (Abu-Mostafa, Magdon-Ismail, and Lin, 2012).

Without active information, machine learning practitioners would never be able to find hypotheses that make good predictions. However, machine learning practitioners regularly find very good hypotheses. The success of machine learning has made numerous companies, such as Google, very wealthy. This success is all contingent on the existence of active information. Therefore, methodological naturalism is false and causes exist that can create active information. The challenge is to describe, identify, and isolate these causes.

2.3 Calculating Complexity (category 2)

Criticism: *Non-naturalism needs a practical method for quantifying whether an effect has a natural cause.*

In the ID metric, complex specified information (CSI), the complexity of an object must be calculated, which is the probability that it occurs through chance and/or necessity (Dembski, 2005). But, this is extremely hard to calculate for real scenarios. We would have to understand something completely in order to calculate its likelihood of occurring through chance and/or necessity, and we cannot understand anything completely.

There are statistical approaches that can estimate complexity for processes based on observed evidence. Scientific theories and laws can also be used to create models to estimate the naturalistic hypothesis. However, there is no general method to calculate the complexity of a given artifact.

Without a general method, it is not possible to know whether an arbitrary artifact is designed without knowing what entity produced it. Yet, if we must know the entity, then this defeats the purpose of ID. ID is supposed to identify intelligent design without reference to the designer.

2.4 Independent Specification (category 2)

Criticism: *Non-naturalism needs a practical method for quantifying whether a pattern can be separated from its instantiation.*

The second half of the CSI metric is specification (Dembski, 2005). Specification is a pattern that describes an event. The pattern must be independent from the event. How do we guarantee the pattern is independent?

Let us assume E is an event, H is the space of possible events, and K is the event's specification. Independence means $P(E|H) = P(E|H, K)$. In other words, the probability of the event does not increase when the specification is included. This requirement is clearly expressed mathematically, but it needs to be worked out in practice. For a given situation and specification, how do we show the specification is independent of the event?

There are some heuristics that can be used, and in certain cases, it is obvious that the specification is independent. For example, alphabet magnets on a refrigerator that spell "buy milk" would be an example of an event whose internal structure is independent of the specification." There is nothing about the magnets' internal structure that makes it any more likely to arrange themselves into a coherent English phrase.

Another example is a written math equation. The mathematical concept represented by the equation is independent from the physical symbols and the ink on paper used to write the equation. We can know the concept is independent from the symbols because the concept does not cause the written equation to come into being.

More work along these lines to develop techniques and guidelines for determining independence is needed.

2.5 Subjective Specification (category 3a)

Criticism: *Non-naturalism needs an objective way to define patterns.*

Specification is often considered a subjective pattern in the ID literature. If it is subjective, how is confirmation bias avoided? Furthermore, if specification is completely subjective, then one pattern is just as good as another. In which case, there is a specification for every event, and the true CSI calculation for everything is zero. Therefore, in order to have a CSI calculation greater than zero, an objective specification is necessary.

Dr. Dembski and Dr. Ewert have both used Kolmogorov complexity as an objective specification (Dembski, 2005; Ewert, Dembski, and Marks II, 2012). Even Dr. Elsberry, within his article criticizing Dr. Dembski's work, uses Kolmogorov complexity to create an alternative to CSI (Elsberry and Shallit, 2011). In general, Dembski describes a specification as a concise description (Dembski, 2005). As Kolmogorov complexity is a way to quantify conciseness, it appears conciseness is one way to make specification objective.

It is important to note that an objective specification is distinct from an independent specification. A specification can be objective without being independent, such as a bitstring that is generated by its shortest possible program instead of randomly. A specification can also be independent without being objective if it is chosen arbitrarily by a being with free will. An example of this is using a source of true randomness to generate a cryptographic key.

2.6 Non-naturalism Needs to Make Money (category 1)

Criticism: *Non-naturalism needs to prove itself through reduction to practice and make money.*

According to Claude Shannon, if everyone else believes in a false probability distribution and you know the true distribution, the difference between the two distributions is worth money. The distance between the chance and necessity distribution and the non-naturalistic distribution is unbounded since the latter allows for the creation of information. Consequently, non-naturalistic theories of science should be making more money than methodological naturalism.

This is a kind of falsification/confirmation by practice. If a theory claims to provide such an extremely different account of reality than the dominant worldview, it should produce significant effects the dominant worldview cannot produce.

2.7 Both Non-Deterministic and Non-Random (category 2)

Criticism: *Non-naturalism needs a clear description of what a non-deterministic and non-random cause looks like.*

Methodological naturalism seeks to explain everything in terms of determinism and randomness. In other words, it describes all of reality using mathematical formulae relating the past to the present. Therefore, a non-naturalistic theory needs to explain reality without using either determinism or randomness. What does it look like to describe reality with mathematical rigor without using determinism or randomness?

The chance and necessity description of reality gave us the physical laws, chemical reactions, and algorithmic processes. All have proven remarkably useful for improving our lives and expanding our abilities.

A non-naturalistic theory of reality will need to also provide the equivalent types of advances. It will require new kinds of laws that take into account purpose and design. There has hardly been any work to develop these new laws. Besides the formulation of ID theory by Dembski, the only other mathematical attempt the author knows of is Hawthorne and Nolan (2006).

On the other hand, it is ironic that though naturalism is described mathematically, naturalism cannot explain the existence of mathematics. For example, the number two exists independent of any physical instantiation. If I were to burn a paper with the number two written on it, I have not destroyed the number two. Additionally, the number two existed before it was ever used by anyone. Another example is the concept of infinity. While infinity is essential to many mathematical concepts, such as calculus and irrational numbers, it cannot have a finite physical existence.

2.8   Real Halting Oracle (category 1)

Criticism: *Non-naturalism entails the existence of uncomputable causes.*

In the computer science discipline of search and optimization, there is a law known as the *No Free Lunch Theorem (NFLT)*. The NFLT states that all search algorithms are exactly equal in performance when averaged over all problems. For example, if we were to make a prediction algorithm, this algorithm would perform no better than random prediction when averaged over all prediction problems. However, the creation of active information violates this law and allows for things such as prediction better than chance.

If an entity can create active information, then it will find a target, on average, better than any search algorithm. This invalidates the NFLT. However, invalidating the NFLT requires calculating Kolmogorov complexity.

To understand why it is necessary to calculate Kolmogorov complexity, we need to look at Occam Learning. *Occam Learning* states that compression leads to prediction in machine learning (Blumer, Ehrenfeucht, Haussler, and Warmuth, 1990). If we can calculate compression, which generally is Kolmogorov complexity, then we can predict it. But, the NFLT states that in general, prediction is not possible. So, if we can calculate Kolmogorov complexity, we can predict and consequently invalidate the NFLT.

Unfortunately, calculating Kolmogorov complexity is uncomputable. But, with a halting oracle, it is trivial to calculate Kolmogorov complexity. To do so, skip all programs that do not halt and check all programs that halt until one produces the correct bitstring.

Pulling this altogether, if active information is created, then halting oracles are real. This is a non-trivial claim, since most computer scientists say halting oracles do not exist.

2.9   Reverse Entropy (category 1)

Criticism: *Non-naturalism violates the second law of thermodynamics.*

The second law of thermodynamics states the entropy of a system is always increasing if there is no external intervention. Entropy is mathematically defined as the tendency of a system to move toward highly probable states. CSI is by definition the movement towards highly improbable states. As CSI increases linearly, the improbability grows exponentially. Consequently, CSI implies an exponential ability to reverse entropy.

This can be seen in a concrete way with CSI. If we measure the CSI for an event X, we have to choose appropriate metrics for complexity and specification. These choices of metrics assume the event X originates from a context where the chance hypothesis is a uniform distribution. X is encoded as a bitstring of finite length, so all X's look something like 0100001110101110.

If we use self-information $I(X)$ for our complexity measure and Kolmogorov complexity $K(X)$ for our specification measure, then self-information can be simply stated as the length of a random bitstring. For example, the randomly generated bitstring 010011 has self-information of 6. Kolmogorov complexity is the shortest, lossless compression of a bitstring. The regular bitstring 111111111 has a much smaller Kolmogorov complexity than the random bitstring 1010001011.

With these metrics, CSI is calculated using the following equation:

$$CSI = I(X) - K(X)$$

For all X of the same length, $I(X)$ will be the same. If the event X exhibits a pattern, then it will have low $K(X)$ and thus high CSI.

We can also define entropy in terms of Kolmogorov complexity (Gell-Mann and Lloyd, 2004).

$$Entropy = K(X)$$

From this we can easily see that CSI is inversely proportional to entropy. As more CSI is created by an intelligent agent, there is a smaller amount of entropy. Therefore, intelligent agents have the ability to reverse entropy through creation of CSI. This relationship does not necessarily mean intelligent agents can create energy out of nothing. The decrease in entropy might be offset by an increase in entropy elsewhere. However, this does not imply the converse.

As Granville Sewell has shown, an increase in entropy in one area does not automatically result in decreased entropy in another area (Sewell, 2001). For example, lighting a match near a gas leak will increase entropy dramatically without decreasing entropy anywhere else.

In order for an increase in entropy to reduce entropy elsewhere, an orderly application of the increasing entropy is required. This happens frequently in intelligently designed mechanisms such as internal combustion engines. Consequently, even if intelligent agency only decreases entropy locally, while increasing entropy globally, the existence of entropy reduction cannot be explained by physical laws.

2.10   Break in Causal Chain (category 1)

Criticism: *Non-naturalism breaks physical causal closure.*

If CSI exists, then it means that at some point the causal chain A → B → C was broken. The complexity part of CSI is the probability that the event happened by chance and/or necessity. Chance and/or necessity (CN) form the causal chain A → B → C of the material world. Deviations from CN break the causal chain. If an event contains CSI, then it cannot be explained by CN and the causal chain has been broken.

Moreland and Craig independently predicted that a non-natural cause must necessarily break the causal chain (Moreland and Craig, 2003, ch. 17).

> This means that when it comes to states of affairs directly produced by agent causes (the hand being raised, life being created), there will be a gap between the state of affairs that existed prior to the effect and the state of affairs that is (or is correlated with) that effect.

2.11   Immaterial Reality (category 1)

Criticism: *Non-naturalism requires that an immaterial reality exists.*

It is a basic truth that something cannot come from nothing. Even the quantum events that pop into existence come from a background probability distribution of possible events (He, Gao, and Cai, 2014). This space of possible events has existence of some sort, even though it may be called "nothing."

As described in the previous section, since non-naturalism implies the breaking of the material causal chain, something must have broken the causal chain. This something cannot be material since it acted contrary to materialism. And, since something cannot come from nothing, it must be a real thing. Consequently, there has to be an immaterial reality.

Furthermore, a counterintuitive result from using Occam's razor is that only the immaterial reality needs to exist. For example, CSI can only give true positives when detecting intelligent agents; it cannot give true negatives. While we can be sure that something is the product of an intelligent agent, we can never know whether something is not the product of an intelligent agent. Consequently, everything can be explained as the product of intelligent agency, and there is no need to use chance and necessity as an explanation.

This does not work the other way around. Chance and necessity cannot explain CSI, so intelligent agency is always a required explanation.

The original statement of William of Occam's famous principle is "Numquam ponenda est pluralitas sine necessitate." Translated this says, "Plurality must never be posited without necessity." In other words, we should never try to explain something with more concepts when we can do so with fewer. If we use this principle strictly, then we have to say matter does not exist. Instead, only intelligent agency exists.

In this case, methodological naturalism is grossly inadequate to rationally explain existence. Furthermore, the immaterial world exists to a greater degree than the material world since it can explain everything that we consider to be material.

The immaterial world is also the source of everything we value in the material world, and more valuable than the material world. We have seen that everything in the material world can be explained by the immaterial world. Since a greater thing cannot be explained by a lesser thing, then the immaterial world is of much greater value insofar as it explains everything we materially value.

2.12 Non-Naturalism Worldview (category 1)

Criticism: *Non-naturalism both contradicts the pervasive modern worldview across all disciplines and requires a new worldview.*

Methodological naturalism is not just restricted to biological history. It has impacted every single field of academia and created new fields. Consequently, non-naturalism has to likewise impact and recreate our worldview.

The standard approach to integrating academic disciplines and non-naturalistic theories, such as theology, is to apply the principles of the discipline to the theory. However, if a non-naturalistic theory is true, then the reverse should occur. This is even more so if the theory is true and more fundamental than the principles of the academic discipline.

However, if a non-naturalistic theory is true, and an academic discipline is true, then a non-naturalistic theory should not contradict the academic discipline. Instead, the academic discipline should be improved by the incorporation of ideas from a non-naturalistic theory.

This means that research and educational institutions that are based on non-naturalistic theories should make more profound contributions to all academic disciplines than those founded upon naturalism. Conversely, the more institutions accept only naturalistic principles, the less significant will be their contributions.

But note that significance is a measure of quality, not quantity. For example, Einstein only wrote a few papers in his career, but he had an enormous impact. So, while naturalism may result in many academic papers, non-naturalism will result in more impactful papers.

2.13   What Makes Intelligent Agents? (category 3a)

Criticism: *Non-naturalism needs to explain the origin of non-naturalistic agency.*

If the following is true:

1. Intelligent agents are embodied, therefore they have a beginning, (see Section 2.1)

2. Intelligent agency cannot result from material processes, (see Sections 2.2 and 2.7–2.10)

then intelligent agents are created by an immaterial being.

This conclusion has a couple of profound implications. First of all, this gives intelligent agents enormous inherent value. In Section 2.11 we saw how the immaterial realm is more valuable than the material realm. Since intelligent agents come from this realm, they also have more value than purely material goods.

Second, insofar as intelligent agents have great inherent value, their creator consequently has even more value. These implications create a value scale in our world. We should value intelligent agents above purely material goods. We should value the creator of intelligent agents above everything else.

3   Conclusion

These numerous criticisms are distinct from the standard criticisms of non-naturalism because they are from a proponent's perspective. As such, they come from a well-developed understanding of non-naturalism. Additionally, these criticisms are constructive. They identify areas where non-naturalism can improve. These criticisms also point out implications that are very counter-intuitive from a naturalistic perspective. This is the sort of criticism that non-naturalistic theories need.

While most of these criticisms target Intelligent Design in particular, their characteristics will be common to most non-naturalistic theories. As defined at the beginning, naturalism is encompassed by all that is algorithmically describable. An algorithm is a Turing machine, which is a realization of all that is mechanically possible.

Non-naturalism, consequently, refers to all that is not algorithmic and not explainable by chance and necessity or mechanism. This means there is no way to formulate a non-naturalistic method and metaphysic that avoids the criticisms presented here.

Conversely, if a theory does avoid all the aforementioned criticisms, it may be a naturalistic theory. The theory is necessarily still a mechanical theory, since it can be encompassed by a Turing machine. For example, a theory that explains all of reality using a god that does not have libertarian free will can still be described by a Turing machine. Thus, such a theory describes reality with a mechanical god.

These criticisms not only point out how non-naturalistic theories can be improved, but also serve as a litmus test for whether a theory counts as non-naturalistic.

References

Abu-Mostafa, Y.S., Magdon-Ismail, M., and Lin, H.T. 2012. *Learning from data.* AMLBook Berlin, Germany.

Beckwith, F. 2010. Intelligent Design and me, part 2: Confessions of a doting Thomist. http://biologos.org/blogs/archive/intelligent-design-and-me-part-2-confessions-of-a-doting-thomist

Blumer, A., Ehrenfeucht, A., Haussler, D., and Warmuth, M.K. 1990. Occam's razor. *Readings in machine learning* pp. 201–204.

Chu-Carroll, M. 2006. Good math/bad math. http://goodmath.blogspot.com/2006/04/qa-roundup-no-free-lunch.html

Dawkins, R. 2008. *The god delusion.* Houghton Mifflin Harcourt.

Dembski, W.A. 2005. Specification: the pattern that signifies intelligence. *Philosophia Christi* 7(2):299–343.

Dembski, W.A. and Marks II, R.J. 2010. The search for a search: Measuring the information cost of higher level search. *JACIII* 14(5):475–486.

Elsberry, W. and Shallit, J. 2011. Information theory, evolutionary computation, and Dembski's "complex specified information". *Synthese* 178(2):237–270.

Ewert, W., Dembski, W.A., and Marks II, R.J. 2012. Algorithmic specified complexity. *Engineering and Metaphysics, Tulsa, OK* .

Gell-Mann, M. and Lloyd, S. 2004. *Effective complexity.* Oxford University Press: Oxford, UK.

Hawthorne, J. and Nolan, D. 2006. What would teleological causation be. *Metaphysical essays* pp. 265–284.

He, D., Gao, D., and Cai, Q.y. 2014. Spontaneous creation of the universe from nothing. *Physical Review D* 89(8):083510.

Melott, A.L. 2002. Intelligent design is creationism in a cheap tuxedo. *Physics Today* 55(6):48.

Moreland, J.P. and Craig, W.L. 2003. *Philosophical foundations for a Christian worldview.* InterVarsity Press.

Sewell, G. 2001. Can anything happen in an open system. *The Mathematical Intelligencer* 23(4):8–10.

Wolpert, D. 2003. William Dembski's treatment of the no free lunch theorems is written in jello. *Mathematical Reviews* .

Part III

Alternatives to Naturalism in Specific Fields

Establishing alternatives to methodological naturalism requires concrete proposals. How should science and other academic fields be engaged if naturalism is removed? These papers provide alternative ways to engage a variety of fields in non-naturalistic ways.

Computer Science: Imagination Sampling

ERIC HOLLOWAY

Baylor University

Abstract

Machine Learning, despite its name, can incorporate an oracle. One common form of oracle interaction is known as *active learning*. Active learning samples $\{x, y\}$ from an oracle for f (the function to be learned). *Imagination sampling* is the converse of active learning. Imagination sampling asks an oracle for hypotheses h from $\mathcal{H}$ (hypothesis space). In this paper imagination sampling is compared with a purely algorithmic approach to determine if oracle interaction outperforms a purely algorithmic approach. The theoretical basis for imagination sampling is developed and illustrated by simulating an oracle.

1 Introduction

The effectiveness of a machine learning algorithm can be measured by the number of samples required to match f (problem) and $h \in \mathcal{H}$ (hypothesis space). One approach to improving machine learning is to incorporate oracle interaction. An oracle is an external source of information. Machine learning approaches that incorporate oracle interaction focus on how oracle interaction significantly improves the sampling of f. This is known as *active learning*, along with a variant known as *guided search* described below. However, there is no research into whether an oracle improves sampling of $\mathcal{H}$.

2 Background

A sub-field of active learning is guided search. In guided search an oracle not only labels data items, but also identifies data items for labeling. Attenberg et al. have shown that a guided search is superior to active learning in the area of website classification (Attenberg and Provost, 2010). It is useful for domains where the target classification has a very small number of instances compared to the general population. Additionally, guided search is good for classifications that are formed from disjunctive subclasses. These are common problems encountered when putting active learning to use (Attenberg and Provost, 2011). A similar approach is oracle guided feature selection for particular classes. This is known as *guided feature labeling* (Attenberg, Melville, and Provost, 2010). Most recently, Attenberg has proposed a gamified system called *beat the machine* (BTM). BTM uses oracles to identify the unknown unknowns of a machine learning model. (See Attenberg, Ipeirotis, and Provost, 2015 for more information.)

Classification model learning is divided into three different areas:

1. Query to classify data: $y = f(x)$

2. Select data to be classified: $x \in \mathcal{D}$

3. Select predictive model: $h \in \mathcal{H}$

There are four different approaches to optimizing these areas:

A. Random

B. Algorithmic

C. Oracle

D. Ground truth

3D is the goal of machine learning—an accurate prediction model. Traditional machine learning is a combination of 2A (random sampling), 3B, and 1D. Semi-supervised learning introduces 1B. The learned model is used to classify further data for learning.

3A is used in the *No Free Lunch Theorem* to characterize the expected performance of machine learning.

Active learning incorporates 1C (oracle labeler) and 2B into traditional machine learning. Guided search adds 2C. Finally, BTM and its predecessor *equivalence querying* (Angluin, 1988) are another form of 2C.

Table 10.1 shows the areas covered in the literature. Most of the combinations have been addressed.

Another approach is for an oracle to sample from $\mathcal{H}$, the hypothesis space. This approach is 3C. Sampling from $\mathcal{H}$ uses the oracle's imagination so this approach is called "imagination sampling."

Oracle sampling of $\mathcal{H}$ is unexplored in the literature. This project investigates whether an oracle improves the sampling of $\mathcal{H}$ by comparing the effectiveness of "imagination sampling" with algorithmic approaches.

Table 10.1: Research grid

	A	B	C	D
1	X	X	X	X
2	X	X	X	
3	X	X		X

3 Testing "Imagination Sampling"

The No Free Lunch Theorem (NFLT) states that all algorithms have exactly the same performance when averaged over all problem domains. While particular algorithms perform better on particular problem domains, it is extremely unlikely to pair the right algorithm with the right domain. Consequently, the NFLT is used in two different ways to test for non-algorithmic learning in the oracle: (1) in terms of the problem domain, and (2) of the learning algorithm.

As stated, it is unlikely a particular algorithm will do well on a randomly selected problem. If imagination sampling performs well on an arbitrary problem domain, then the oracle is highly likely to have a non-algorithmic learning ability. The oracle cannot have information about the dataset that the algorithms do not have.[1]

Similarly, for a particular problem, it is unlikely a randomly selected algorithm will do well. The learning problem is constructed so algorithmic learning cannot perform better than random sampling that is averaged over many problems. This is a No Free Lunch (NFL) construction.

In either case, if the oracle performs better than algorithmic approaches, this shows imagination sampling is generally better than purely algorithmic learning.

The NFL construction is a hypothesis space that shatters every dataset. A hypothesis space shatters a dataset when it can represent any possible labeling of items in the dataset. Algorithmic learning based on in-sample error (E_{in}) is not possible in this hypothesis space, since a hypothesis with zero in-sample error can always be found. However, learning based on compression is possible and will be covered.

The oracle is compared to two classes of algorithms. The first class samples hypotheses over the entire hypothesis space. The second class learns using a learnable subset. The subset is learnable because it does not cover all possibilities and in-sample error cannot be completely minimized. The oracle should be more effective than the first class, and may be more effective than the second class. Effectiveness is measured by the lowest out-of-sample error, E_{out}, obtained.

A second issue is identifying when the oracle has found a good hypothesis. Since

---

[1] This sounds like a contradiction since an oracle is defined as an external source of information. If the oracle does not have information, then it is not a source of information. However, some oracles can create information. Algorithms cannot create information. So, relying on an oracle that does not initially have information can still be useful if the oracle can create information.

the NFL construction shatters every dataset, overfitting is more likely to occur than generalization. This means the hypothesis cannot be rated in-sample using E_{in} and E_{test}, the test dataset. Instead, the hypothesis must be rated by its conciseness. Conciseness is measured by Kolmogorov complexity (KC).

This paper uses KC to identify when the oracle has found a good hypothesis. Only an upper bound for KC can be calculated. Consequently, a gradient-based approach is used with the upper bound to identify a good hypothesis.

4   Representation

As discussed previously, an NFL construction for the learning problem is a hypothesis space that shatters every dataset. To shatter a dataset, we need a hypothesis space that can represent all possible classifications for a dataset. The multibox is a hypothesis space that represents all possible classifications for any dataset, and can shatter any dataset. This makes the multibox an NFL construction. The multibox is defined after the problem domain is described.

4.1   Problem Domain

The problem domain for this project is a 2D discrete grid. The target function f is a particular classification of the cells, such as in Figure 10.1. The classification in the image is represented by white and black cells. A hypothesis h from the hypothesis space $\mathcal{H}$ is a classification region represented by light and dark gray cells. The classification region does not have to classify the entire problem domain.

4.2   Multibox Definition

A multibox is a set of n coordinate 5-tuples. The fifth element is the box's binary classification. Each tuple defines an axis-aligned, rectangular box. The coordinates are integers. The set of all boxes is MB. The set of boxes in the hypothesis is MB_g.

$$MG_g = \{Z, Z, Z, Z, 0 \text{ or } 1\}^M$$

The parameters of each box are referenced by a subscript, box_{0-4} with box_4 as the 0 or 1 classification.

The oracle places multiple axis-aligned, rectangular boxes on the data to form a hypothesis. Each box classifies samples according to a single classification. For example, a box that classifies samples as "ones" will classify all contained samples as ones. If boxes overlap, samples are classified according to the last placed box. The classification algorithm is shown in Algorithm 1.

We can see the multibox represents any classification if each box is the size of an individual cell in Figure 10.1. This means the multibox can shatter all datasets, and

Figure 10.1: Problem domain with an example target function (white and pure black) and classification region (light gray and dark gray).

algorithmic learning is not possible. Algorithmic learning based on error minimization cannot happen because an error can always be taken to zero. Learning through compression cannot happen as compression is not computable in general.

Due to the NFLT the odds of a learning approach performing well on a particular problem are too small to happen by chance. This is still true even if most or all real world problems are learnable because the NFLT applies to learning approaches as well as problems. If an approach does well on a decent number of different datasets, and it is selected independently from the problem domain, then it is superior to algorithmic learning.

Consequently, if experiments show imagination sampling is superior to algorithmic approaches for multibox classification and other classification tasks then it is superior to algorithmic learning in general.

Algorithm 1 Multibox classifier

1: **procedure** CLASSIFY(x)
2:     $result \leftarrow$ NULL
3:     **for all** $box \in MB_g$ **do**
4:         **if** $x \in box$ **then**
5:             $result \leftarrow box_4$
6:     **return** $result$

---

5   Methodology

To test whether imagination sampling is superior to algorithmic techniques, the oracle is compared to two classes of algorithms. The first class learns over the entire hypothesis space. The second class learns a learnable subset.

5.1   First Class of Algorithm

The algorithms in the first class are random box placement and guided box placement. The random approach places boxes randomly until it has placed M boxes (see Algorithm 2). Each box's classification is assigned randomly.

Algorithm 2 Random placement

1: **procedure** RANDOMPLACEMENT(X, M)
2:     $MB_g \leftarrow$ LIST
3:     **for all** *iteration* $\in 1$ to M **do**
4:         $x_{00} \leftarrow$ UNIFORM$(\min_i X_{0i}, \max_i X_{0i})$
5:         $x_{01} \leftarrow$ UNIFORM$(\min_i X_{0i}, \max_i X_{0i})$
6:         $x_{10} \leftarrow$ UNIFORM$(\min_i X_{1i}, \max_i X_{1i})$
7:         $x_{11} \leftarrow$ UNIFORM$(\min_i X_{1i}, \max_i X_{1i})$
8:         $class \leftarrow$ CHOOSE$(0, 1)$
9:         $MB_g.$APPEND$(\{x_{00}, x_{01}, x_{10}, x_{11}, class\})$
10:    **return** MB_g

The guided box placement algorithm is initiated with the random algorithm by placing N boxes. Then, it selects a subset M that maximizes the significance score in Algorithm 3.

The first term, *accuracy*, prioritizes boxes that contain an unlikely number of samples. The second term, *correctness*, prioritizes boxes that are likely to correctly classify a sample.

α and β are tunable parameters between zero and one. When both parameters are zero, guided placement reduces to random placement. Two parameters are used instead of just setting $\beta = 1 - \alpha$, otherwise it is not possible to set both to zero and achieve a random box placement. Consequently, both the random and guided algorithms can be described by Algorithm 4.

5.2   Second Class of Algorithm

The second class of algorithms uses a learnable subset of $\mathcal{H}$. The algorithm in the second class of algorithms is the Set Cover Machine (SCM) algorithm (Marchand and Taylor, 2003). The SCM algorithm finds a minimal multibox of square boxes that cover all of one class of samples. A regularization parameter p penalizes classification

---

Algorithm 3 Significance scoring

1: **procedure** $\textsc{Significance}(X, Y, box, \alpha, \beta)$
2:     $X_{width} \leftarrow \max_i X_{0i} - \min_i X_{0i}$
3:     $X_{height} \leftarrow \max_i X_{1i} - \min_i X_{1i}$
4:     $X_{area} \leftarrow X_{width} X_{height}$
5:     $\lambda \leftarrow \frac{|X|}{X_{area}}$
6:     $box_{width} \leftarrow box_1 - box_0$
7:     $box_{height} \leftarrow box_3 - box_2$
8:     $box_{area} \leftarrow \textsc{abs}(box_{width} box_{height})$
9:     $k \leftarrow \frac{|x \in box|}{box_{area}}$
10:     $probability \leftarrow \textsc{poisson}(k, \lambda)$
11:     $accuracy \leftarrow -\alpha \textsc{log}(probability)$
12:     $probability \leftarrow \textsc{abs}(1 - box_4 - \frac{\sum_{y \in box} y}{|y \in box|})$
13:     $correctness \leftarrow -\beta \textsc{log}(probability)$
14:     **return** $accuracy + correctness$

Algorithm 4 Generalized box placement

1: **procedure** $\textsc{BoxPlacement}(X, Y, N, M, \alpha, \beta)$
2:     $MB_g \leftarrow \textsc{RandomPlacement}(X, N)$
3:     $scores \leftarrow \textsc{LookupTable}$
4:     **for all** $box \in MB_g$ **do**
5:         $scores[\textsc{Significance}(X, Y, box, \alpha, \beta)] \leftarrow box$
6:     $\textsc{SortDescending}(scores)$
7:     $MB_g \leftarrow scores.\textsc{values}[1 \dots M]$
8:     **return** MB_g

---

error in the SCM. p can range from zero to ∞. In this experiment, $p = \infty$ so there is no misclassification error. Other values were tried but did not noticeably improve accuracy, and they greatly increased computation time.

The standard SCM algorithm classifies samples outside of the set cover as the alternative class. The SCM algorithm is modified for this experiment to only classify the samples covered by the multibox and ignore samples outside the multibox. In this way, it is similar to a deterministic version of generalized box placement in Algorithm 3.

The standard SCM algorithm has an $O(N^4)$ complexity. To improve runtime, the sample count for the SCM is reduced to $\sqrt{N}$, while the other approaches still use N samples.

6 Experiment

The dataset comes from a BNP insurance competition on Kaggle.com. and has 130 anonymized features and 130k samples. Anonymization keeps the oracle from having access to domain knowledge.

The four approaches are trained and validated across a range of sample sizes for E_{in} and E_{test}, from 100 to 500 samples using 50 sample increments. A separate set of 1000 samples is used for calculating E_{out}. The oracle's multibox placement is compared to the algorithmic approaches using E_{out} in a batch after all experiments are completed. E_{out} is calculated using Algorithm 1. s_C represents the E_{out} samples covered by the hypothesis.

$$E_{out} = \sum_{x,y \in s_C} \text{abs}\left(\frac{y - \text{classify}(x)}{|s_C|}\right)$$

1. The samples are drawn randomly without replacement from the dataset to create the training set. The samples are reduced to remove samples that are missing data and are further reduced so there are an equal number of both classes. This can result in a sample set containing much less than the initial amount. For instance, if 100 samples are initially drawn, cleaning and equalizing can leave only 30 samples.

2. The samples are preprocessed to be centered, normalized, and whitened. The parameters for the preprocessing are recorded for use on the E_{test} and E_{out} samples.

3. The two algorithmic multibox hypotheses are generated with $M = 5$, meaning each hypothesis consists of 5 boxes. This number is selected to be small so the resulting hypothesis has a low complexity. For the guided multibox selection, $N = 100$. This means 100 boxes are generated randomly. Then the 5 best boxes

are selected for the hypothesis, as shown in Algorithm 4. The guided multibox algorithm has parameters $\alpha = 0.5$ and $\beta = 0.5$ in Algorithm 3.

4. The oracle (human user) visually selects a set of boxes it thinks will create a good classification hypothesis. Figure 10.2 is an example MB_g created by an oracle.

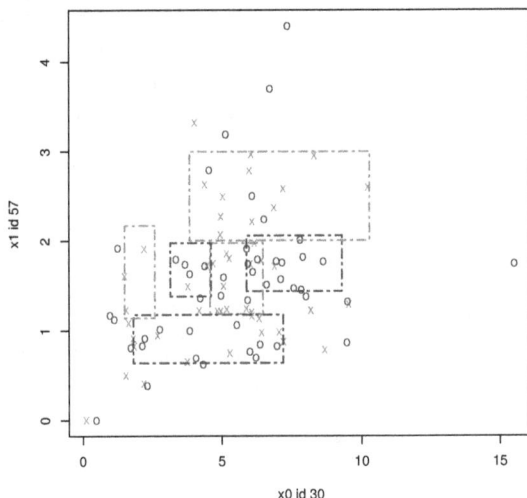

Figure 10.2: Example of MB_g generated by oracle.

5. Each hypothesis is then evaluated for classification error on a validation and test dataset. Both datasets use different samples, which are also cleaned, equalized, and preprocessed. The training dataset values are used for preprocessing.

7   Imagination Sampling Results

The experiment is repeated 351 times and is carried out across sample sizes from 100 to 500, with 50 sample increments. In each experiment, the approaches are compared based on accuracy, $1 - E_{out}$. The outcome of an experiment identifies which hypothesis of the four approaches has the best accuracy. The approach with the highest accuracy wins the experiment.

The overall results as well as wins, min, mean, and max accuracy for each sample size are shown in Table 10.2. Some numbers are truncated to fit the table. As the results show, the oracle performs the best.

Table 10.2: Results from 351 experiments.

SCM	All	100	150	200	250	300	350	400	450	500
Wins	91	**12**	10	**12**	10	11	**12**	6	9	9
Max	1	.82	.64	.75	.83	.66	1	.70	.59	.60
Mean	.50	.51	.51	.50	.51	.48	.50	.48	.48	.50
Min	0	.33	.40	.14	.33	.22	.33	0	0	.23
Oracle	**All**	100	150	200	250	300	350	400	450	500
Wins	**114**	8	10	9	**11**	**16**	**12**	**18**	**15**	**15**
Max	.75	.60	.58	.63	.75	.70	.61	.62	.62	.60
Mean	.50	.49	.49	.49	.51	.50	.50	.51	.50	.51
Min	.33	.36	.33	.37	.42	.40	.39	.43	.41	.38
Rand	**All**	100	150	200	250	300	350	400	450	500
Wins	71	10	7	8	7	8	8	4	9	10
Max	1	.75	.65	1	1	.75	.64	.69	.66	.66
Mean	.49	.50	.50	.51	.49	.50	.47	.48	.50	.50
Min	0	.32	.33	.33	0	0	0	.34	.33	.39
Guided	**All**	100	150	200	250	300	350	400	450	500
Wins	75	9	**12**	10	**11**	4	7	11	6	5
Max	.74	.63	.67	.74	.55	.58	.52	.55	.54	.56
Mean	.49	.49	.50	.51	.49	.48	.49	.49	.49	.49
Min	.31	.33	.40	.36	.34	.32	.41	.31	.40	.45

The interesting trend in the results is that as the number of samples increases, the oracle's performance improves relative to the other algorithms. It is also evident that, in general, none of the approaches did very well.

8   Identifying Good Oracle Hypotheses

Using the multibox model means E_{in} and E_{test} cannot be used to evaluate hypothesis accuracy. Since the model shatters every dataset, a hypothesis can always be found that makes $E_{in} = 0$ and $E_{test} = 0$. We cannot trust these error metrics to identify a good hypothesis.

Instead, we use algorithmic complexity theory to identify a good hypothesis. Identifying a good hypothesis is a compression problem (Kearns and Vazirani, 1994). But, calculating an arbitrary bitstring's optimum compression is undecidable (Kolmogorov, 1998). However, an upper bound on compression can be calculated, and consequently, an oracle's good hypotheses can be identified by using the upper bound with a gradient approach.

To derive the upper bound, we must first define the problem domain. The problem domain is finite and discrete. There are two classes and an equal count of both classes.

- $\mathcal{A}$ is the complete sample space. For example, the problem domain is a 2D grid. Each cell in the grid is a sample. In this case, $\mathcal{A}$ is all of the grid cells.

- h is the hypothesis being examined. $\mathcal{H}$ is the hypothesis space. $\mathcal{H}$ represents all classifications on the dataset. $diversity(\mathcal{H})$ counts the unique classifications represented by $\mathcal{H}$. Thus, $diversity(\mathcal{H}) = 2^{|\mathcal{A}|}$ and $|\mathcal{H}| \geq 2^{|\mathcal{A}|}$.

- h does not necessarily classify every cell in $\mathcal{A}$. The set of cells that are classified by h is $\mathcal{C}$.

- The set of samples is $\mathcal{S}$. The subset of $\mathcal{S}$ in $\mathcal{C}$ is $s_{\mathcal{C}} \in \mathcal{S}$.

- The Kolmogorov complexity of h is $K(h)$. $\mathcal{H}_k$ is the set of hypotheses where $K(h) = k$. As such, $diversity(\mathcal{H}_k) \leq 2^k$ and $|\mathcal{H}_k| = 2^k$. For a classification region $\mathcal{C}$ there are $2^{|\mathcal{C}|}$ different classifications. $\mathcal{H}_k$ can only describe, at most, 2^k different classifications. So $\mathcal{H}_k$ covers, at most, $2^{k-|\mathcal{C}|}$ possible classifications.

- If the hypothesis correctly classifies all samples, then the conditional algorithmic complexity of the samples given the hypothesis is zero, $K(s_{\mathcal{C}}|h) = 0$. However, if there are misclassifications, then the conditional complexity is non-zero because extra information is needed to describe the misclassified samples. The combination of both hypothesis complexity and sample conditional complexity is *classification complexity*. The expression for classification complexity is $CC(s_{\mathcal{C}}, h) = K(h) + K(s_{\mathcal{C}}|h)$.

- The calculable upper limit for classification complexity with no misclassifications is $D(h) \geq CC(s_{\mathcal{C}}, h)$.

In this analysis, $\mathcal{H}$ is the multibox classifier. (See Algorithm 1.) For an $h \in \mathcal{H}$, $|h|$ is the number of boxes in h. Since each box in h can only have a single classification, then $diversity(h) \leq 2^{K(h)} \leq 2^{|h|}$.

Note, $K(h) \leq |h|$ because there can be a shorter description of h than to enumerate all the boxes. As an example, the boxes form an infinitely long diagonal line. The equation for the line has finite Kolmogorov complexity. The enumeration of the boxes is an infinitely long bitstring. Thus, trivially $K(h) < |h| = \infty$.

As an upper bound on $CC(s_{\mathcal{C}}, h)$, we have $D_1(h) = |h| - \sum_{box \in h} log_2(1 - E_{in}^{box}) * |box|$. The notation $|box|$ counts how many samples from $\mathcal{S}$ are in the box. If all boxes have $E_{in}^{box} = 0$, then there is no need for the second term, and $CC(s_{\mathcal{C}}, h) = K(h)$. However, if the boxes do have an $E_{in}^{box} > 0$, there are a couple of key cases to consider.

1. If $E_{in}^{box} = \frac{1}{2}$, then the box has a 50/50 chance of correct classification. If we go back to our definition of $\mathcal{H}_k$, we see it can describe, at most, $2^{k-|\mathcal{C}|}$ of the

classifications in $\mathcal{C}$. For our subset of samples $s_{\mathcal{C}}$, $\mathcal{H}_k$ can describe $2^{k-|s_{\mathcal{C}}|}$ classifications. If $k = |s_{\mathcal{C}}|$, then $2^{k-|s_{\mathcal{C}}|} = 1$. This means $\mathcal{H}_{k=|s_{\mathcal{C}}|}$ can correctly classify any set of samples of that size and cannot generalize. A hypothesis that cannot generalize has a 50/50 chance of correct classification. Consequently, if $E_{in}^{box} = \frac{1}{2}$, then this is equivalent to $\mathcal{H}_{k=|s_{\mathcal{C}}|}$ for the samples in the box.

2. If $E_{in}^{box} = 1$, then this box is not an acceptable classifier. However, the classification of the box cannot always be changed to turn it into a good classifier. In this case, there are only two classifications, so the box can be fixed. But in general, there are an unlimited number of classifications. In the unlimited case, a box that misclassifies everything cannot be fixed to provide a good classifier.

All these criteria are met by using $log_2(1 - E_{in}^{box}) * |box|$ for the second term.
If $E_{in}^{box} = 0$, $D_1(box) = 1 - log_2(1) * |box| = 1$.
If $E_{in}^{box} = \frac{1}{2}$, $D_1(box) = 1 - log_2(\frac{1}{2}) * |box| = 1 + |box|$.
If $E_{in}^{box} = 1$, $D_1(box) = 1 - log_2(0) * |box| = \infty$.

With a definition of the upper bound on imagination sampling complexity, $D_1(h)$, we need a metric to measure how well a particular classification will perform. In the following discussion, we assume h correctly classifies $s_{\mathcal{C}}$ for simplicity of notation. If h correctly classifies, then $CC(s_{\mathcal{C}}, h) = K(h)$. Additionally, k is used interchangeably with $K(h)$.

For this metric, we need a measure that

1. becomes 0 as $K(h) \to \infty$

2. becomes $\frac{1}{2}$ as $K(h) \to |s_{\mathcal{C}}|$

3. becomes 1 as $K(h) \to 0$

all as $|s_{\mathcal{C}}| \to |\mathcal{C}|$.

We measure the proportion of classifications by $\mathcal{H}_k$ on $s_{\mathcal{C}}$ by $2^{k-|s_{\mathcal{C}}|}$. Therefore, we can define an accuracy metric that follows these criteria.

$$\text{acc}(s_{\mathcal{C}}, \mathcal{C}, K(h)) \leq 2^{\frac{|s_{\mathcal{C}}|-K(h)}{|\mathcal{C}|}-1}$$

If $K(h) \to 0$, then $\text{acc}(s_{\mathcal{C}}, \mathcal{C}, K(h)) \leq 2^{\frac{|s_{\mathcal{C}}|}{|\mathcal{C}|}-1} \to 1$.
If $K(h) = s_{\mathcal{C}}$, then $\text{acc}(s_{\mathcal{C}}, \mathcal{C}, K(h)) \leq 2^{-1} = \frac{1}{2}$.
If $K(h) \to \infty$, then $\text{acc}(s_{\mathcal{C}}, \mathcal{C}, K(h)) \leq 2^{-\infty} \to 0$.
These trends are illustrated in Figure 10.3.

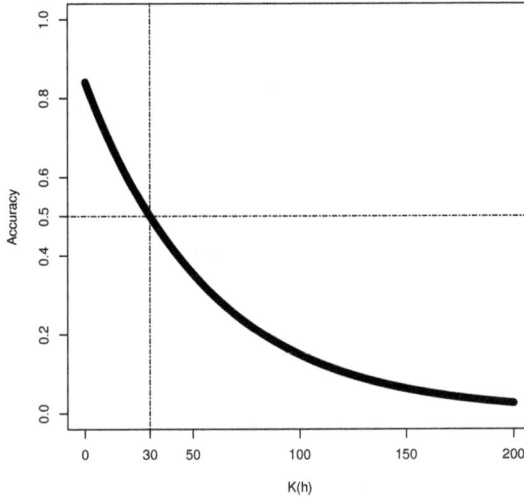

Figure 10.3: Graph of acc function for $\frac{s_{\mathcal{C}}}{\mathcal{C}} = \frac{30}{40}$.

The next item is to invert acc. In other words, given E_{out} and $s_{\mathcal{C}}$, can we find $K(h)$? The answer is: sort of.

We set $acc(s_{\mathcal{C}}, \mathcal{C}, K(h)) = 1 - E_{out}$ and solve for $K(h)$ to get our $acc^{-1}(s_{\mathcal{C}}, \mathcal{C}, E_{out})$ function:

$$2^{\frac{|s_{\mathcal{C}}| - K(h)}{|\mathcal{C}|} - 1} \geq 1 - E_{out}$$

$$\frac{|s_{\mathcal{C}}| - K(h)}{|\mathcal{C}|} - 1 \geq \log_2(1 - E_{out})$$

$$\frac{|s_{\mathcal{C}}| - K(h)}{|\mathcal{C}|} \geq \log_2(1 - E_{out}) + 1$$

$$|s_{\mathcal{C}}| - K(h) \geq |\mathcal{C}|(\log_2(1 - E_{out}) + 1)$$

$$K(h) \leq |s_{\mathcal{C}}| - |\mathcal{C}|(\log_2(1 - E_{out}) + 1) = D_2(h)$$

This metric falls apart if $E_{out} = 0$ or $\frac{|s_{\mathcal{C}}|}{|\mathcal{C}|}$ is too small because $K(h)$ becomes negative. A negative Kolmogorov complexity does not make sense. This discrepancy is probably due to $acc(s_{\mathcal{C}}, \mathcal{C}, K(h))$ having an asymptote of 1, but never reaching 1. Setting $E_{out} = 1$ assumes the asymptote is reached. The other issue is the hidden constant in Kolmogorov complexity, which is not addressed in these equations.

The acc metric does compare favorably with the equation boundaries for Occam Learning from Blumer, Ehrenfeucht, Haussler, and Warmuth (1987). The following equations show the equivalencies with acc.

- n is $K(h)$

- m is $|s_{\mathcal{C}}|$

- ϵ is E_{out}

The parameters $0 \leq \alpha < 1$ and $c \geq 1$ are parameters that specify a particular Occam algorithm.

$$n^c m^\alpha \ln(2) \leq -\frac{1}{2} m \ln(1 - \epsilon)$$

With $\alpha = 0$ and $c = 1$, we can see the boundary conditions are similar to acc. The exception is for $\epsilon = \frac{1}{2}$, which is more stringent than acc.

$$\epsilon = 0 : n \leq \frac{-\frac{1}{2} m \ln(1)}{\ln(2)} = 0$$

$$\epsilon = \frac{1}{2} : n \leq \frac{-\frac{1}{2} m \ln(\frac{1}{2})}{\ln(2)} = \frac{1}{2} m$$

$$\epsilon = 1 : n \leq \frac{-\frac{1}{2} m \ln(0)}{\ln(2)} = \infty$$

The final step in defining the theory of imagination sampling is to find out when we've converged on a good hypothesis. While it is not possible to know if we've found the optimum compression, we can at least measure our progress toward local optima.

If we have found the optimum compression, then $\frac{\Delta K(h)}{\Delta |s_{\mathcal{C}}|} = 0$. Since there is no closed form, or any form, of formula for $K(h)$ the best way we can find the derivative is empirical observation. We want to observe that as $|s_{\mathcal{C}}|$ increases, $K(h)$ remains constant. To estimate $K(h)$ we use validation to get an E_{out} score, and $acc^{-1}(s_{\mathcal{C}}, E_{out})$ to estimate $K(h)$.

Alternatively, we know that since $K(h) \leq D(h)$, then $\frac{\Delta K(h)}{\Delta |s_{\mathcal{C}}|} \leq \frac{\Delta D(h)}{\Delta |s_{\mathcal{C}}|}$ for a large enough Δ. Therefore, if $\frac{\Delta D(h)}{\Delta |s_{\mathcal{C}}|} = 0$, then $\frac{\Delta K(h)}{\Delta |s_{\mathcal{C}}|} = 0$. The intuitive reason for this is if $D(h)$ is constant, then eventually $K(h)$ must become constant. $K(h)$ will not decrease as the number of samples increases.

Once we've empirically solved for $\frac{\Delta K(h)}{\Delta |s_{\mathcal{C}}|} = 0$, and found the optimum h, we will see $E_{out} \to 0$ as both $|\mathcal{C}| \to \infty$ and $|s_{\mathcal{C}}| \to \infty$. This is because in $acc(s_{\mathcal{C}}, \mathcal{C}, k)$, $|s_{\mathcal{C}}|$ and $|\mathcal{C}|$ will grow indefinitely as $K(h)$ remains constant. Thus, $\lim_{|s_{\mathcal{C}}| \to \infty, |\mathcal{C}| \to \infty} \frac{|s_{\mathcal{C}}| - k}{|\mathcal{C}|} = 1$.

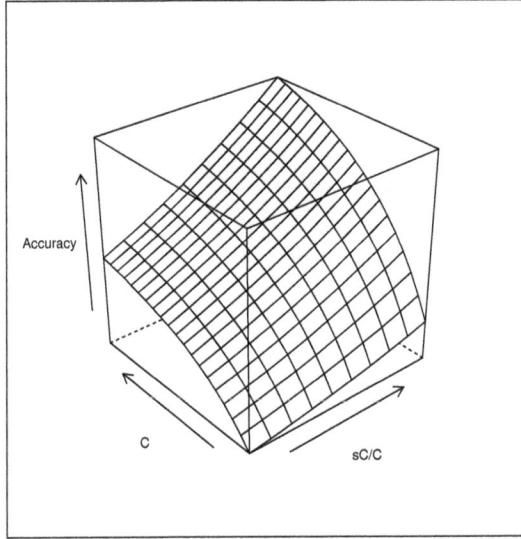

Figure 10.4: Keeping K(h) constant causes accuracy to increase as number of samples increase.

Of course, in a finite, discrete realm both $s_\mathcal{C}$ and $\mathcal{C}$ are bounded by $|\mathcal{A}|$. But as $|\mathcal{A}|$ is enlarged we will see $E_{out} \to 0$.

9   Simulated Oracle Experiment

To test this gradient-based approach with imagination sampling, a simulation of an oracle is used. In the problem, the target function is a tilted rectangle, as exemplified in Figure 10.5. The domain is a 100x100 grid making the rectangle pixelated.

The question is how to simulate a non-algorithmic oracle with an algorithm. Such a simulation seems to be a contradiction in terms. However, the benefit the oracle provides is the ability to infer the target function. To simulate the oracle, we use a learning algorithm that already has the correct class of target function, in this case a tilted rectangle. This learning algorithm is named an Occam learner. Figure 10.5 is the rectangle that the algorithm learned.

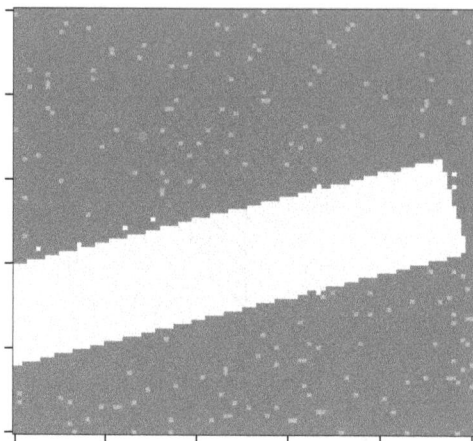

Figure 10.5: Function learned from training samples by Occam learner.

The learned rectangle is not used to calculate the value for $D(h)$. If we used the rectangle, $D(h)$ would give a complexity very close to $K(h)$. But without prior knowledge, we will not be able to calculate $K(h)$ for the oracle's hypothesis. The multibox classifier and E_{test} are used to calculate $D(h)$ instead. The goal is to see if the $D(h)$ metrics are reliable guides for identifying good oracle hypotheses.

The rectangular region is turned into a multibox, with one box for every cell in the rectangular region. $D_1(h)$ is calculated from this multibox hypothesis. $D_1(h)$ is not monotonic as its size will vary based on the samples used to construct the rectangular region. $K(h)$ is also estimated using E_{test} with $D_2(h)$.

The $D(h)$ metrics are calculated for hypotheses learned on different sample sizes. Once a large enough size range has been covered, the gradients $\frac{\Delta D(h)}{\Delta s_c}$ are calculated. The gradient technique is successful if it reliably identifies hypotheses that are highly accurate. To find a good region, we look for areas where at least one of the two gradients show a valley while the other gradient is negative. On the other hand, if the gradient cannot dependably identify accurate hypotheses, then it is not a useful technique. The graph in Figure 10.6 shows that the gradient technique can reliably identify high accuracy hypotheses with one anomaly.

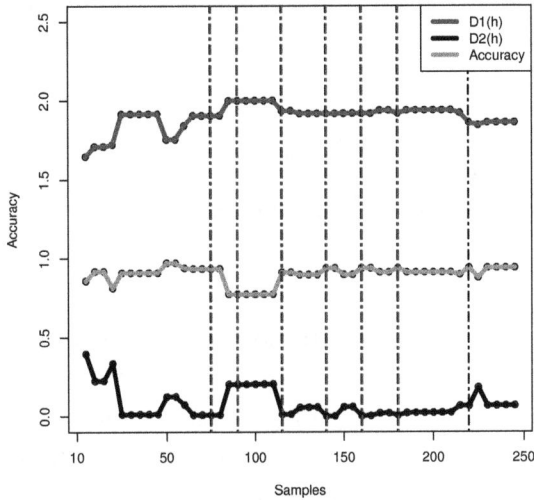

Figure 10.6: Example of the gradient-based approach working effectively. The vertical lines mark points where the gradient is zero and the second derivative is positive. The shade denotes which estimate of $K(h)$ is used to calculate the gradient. $D_1(h)$ is the top line, $D_2(h)$ is the bottom line, and the accuracy is the middle line. Scales are not given for the estimates as the important aspect of the estimates is the gradient.

The gradient technique does not guarantee the global best accuracy. See the second gradient marker from the left in Figure 10.6. While it looks like the marking is in error since it appears that both $K(h)$ estimates are peaks, there are actually slight depressions in the peaks. This shows the gradient method can only guarantee local optima, which may be very local.

10 Empirical Gradient-Based Approach Results

The gradient-based approach is tested using the results from the first experiment.

There are nine different sample sizes ranging from 100 to 500 samples in increments of 50 samples. Each increment is the addition of new samples to the previous set. The actual sample sizes are smaller due to the cleaning and equalizing processes.

The range of samples is tested on a pair of variables, which form the x,y coordinates for the scatterplot. An example of the scatterplot is in Figure 10.2. There are 37 variable pairs in the results.

A particular experiment is identified by the sample size and variable pair. The gradient is calculated over a range of sample sizes for a variable pair. The gradient is then calculated with both the D_1 and D_2 complexity metrics.

The gradient-based approach looks for three consecutive experiments that meet these criteria:

1. s_C increases across all experiments

2. A $D(h)$ metric decreases and then increases

3. The other $D(h)$ metric is not increasing

If these conditions are met, then the gradient has hit a minimal point. When this happens, E_{out} is at a minimal, or has a negative gradient.

There are 259 experiments that can potentially meet the gradient criteria. We cannot know whether the first and last sample sizes are at minimal points so they are excluded. There are 92 (36%) experiments where the E_{out} is at a minimal point leaving only 10 experiments that meet the criteria. Five of the 10 (50%) have an E_{out} at a minimal. The gradient-based approach boosts the accuracy of identifying minimal E_{out} by 14%. The mean accuracy of the 10 experiments is 0.52 and the median is 0.54, both higher than the mean of all the oracle's hypotheses as well as the algorithmically generated hypotheses. However, the p-values for two sided t-tests on these results are 0.27 and 0.24, respectively. Thus, the results are not statistically significant.

11   Conclusion

The purpose of this project is to demonstrate that oracles can generate better hypotheses when compared to algorithmic approaches. To test the oracles' performance against algorithms, an algorithmically unlearnable classification model is used. The classification model, multiboxes, shatters all datasets. This means a good hypothesis must be selected based on compression, but finding a good compression is an undecidable problem. Consequently, due to the No Free Lunch Theorem, no multibox learning algorithm will do better than random sampling.

A dataset with anonymized features is chosen so the oracle does not have access to domain knowledge. The oracle outperforms the tested algorithmic approaches on the anonymized dataset by 114-to-91 successes. Due to the improbability of this result given the NFLT, it shows that the oracle has a non-algorithmic learning capability and can out-perform algorithmic learning in general.

Furthermore, a gradient-based approach for identifying good hypotheses is derived from the theory of imagination sampling. The theory defines how the accuracy of an oracle's hypothesis is based on hypothesis complexity. The complexity of the oracle's hypothesis is not directly calculable but can be estimated with an upper bound. The gradient-based approach is used on the upper bound to identify when a minimal complexity has been found. This minimal complexity identifies a good hypothesis.

The gradient-based approach works with a simulated oracle. The approach is also tested on the results from the initial experiment. It boosts identification of good hypotheses by 14% and improves the mean and median hypothesis accuracy to 0.52 and 0.54. However, the results are not statistically significant, with p-values of 0.27 and 0.24 using the t-test.

References

Angluin, D. 1988. Queries and concept learning. *Machine learning* 2(4):319–342.

Attenberg, J., Ipeirotis, P., and Provost, F. 2015. Beat the machine: Challenging humans to find a predictive model's "unknown unknowns". *Journal of Data and Information Quality (JDIQ)* 6(1).

Attenberg, J., Melville, P., and Provost, F. 2010. Guided feature labeling for budget-sensitive learning under extreme class imbalance. *ICML Workshop on Budgeted Learning* .

Attenberg, J. and Provost, F. 2010. Why label when you can search?: alternatives to active learning for applying human resources to build classification models under extreme class imbalance. In *Proceedings of the 16th ACM SIGKDD international conference on Knowledge discovery and data mining*, pp. 423–432, ACM.

Attenberg, J. and Provost, F. 2011. Inactive learning?: difficulties employing active learning in practice. *ACM SIGKDD Explorations Newsletter* 12(2):36–41.

Blumer, A., Ehrenfeucht, A., Haussler, D., and Warmuth, M.K. 1987. Occam's razor. *Information Processing Letters* 24:377–380.

Kearns, M.J. and Vazirani, U.V. 1994. *An Introduction to Computational Learning Theory*. Massachusetts Institute of Technology.

Kolmogorov, A.N. 1998. On tables of random numbers. *Theoretical Computer Science* 207:387–395.

Marchand, M. and Taylor, J.S. 2003. The set covering machine. *The Journal of Machine Learning Research* 3:723–746.

11 || Linguistics: Grammatical Relations

Noel Rude

Abstract

Human language can be studied bottom-up (corpus linguistics, neurolinguistics) and top-down (via conscious data creation and introspection as to grammaticality). Creativity in language hinges on both law and liberty, on the freedom of the will and constraints thereof. This paper focuses on the role of agency in language, and how our ability to learn and understand language is based not primarily on shared mechanics but rather agency-oriented concepts that we cannot not know.

1 Introduction

Fifty or sixty years ago there actually were linguists who tried to predict the probability of what would come next in a written or recorded text based on what had gone before. Language has its reoccurring and predictable patterns but, of course, one deals here with probabilities, not absolutes. Take any sentence (from here or elsewhere) and do a search on the Internet. I predict you will not find an exact duplicate even in that vast sea.

Leonard Bloomfield, feeling the behaviorism of B. F. Skinner, presided over a mid-twentieth century American linguistics that saw any reference to meaning as unscientific (Bloomfield, 1933). Noam Chomsky famously took Skinner to task (Chomsky, 1959), but nevertheless persevered in the effort to describe grammar without reference to meaning or function.

How then should we build a theory of grammar? From the bottom up based on the structure of what is actually said? Or from the top down deduced from what we know we meant? Why not both ways?

And this is exactly what the generativists of the Sixties and Seventies did—but not without pain and rancor (Harris, 1993; Newmeyer, 1996; Levine and Postal, 2004). The generativists at MIT, as I said, excluded meaning and function from their analyses, but a rebel generative semantics made semantics central. There also emerged a typological/functional school inspired by Roman Jakobson, Joseph Greenberg, and others. The rebels published in edited books subverting the peer review of the journals. [1] Most alternatives to the structuralism of MIT now confess as practitioners of "cognitive linguistics" (Croft and Cruse, 2004; Evans and Green, 2006), and what this seems to be is a quest to prove that the mind equals the brain (Lakoff and Johnson, 1999). But locating human language amid the neurons has not progressed as promised (Pinker, 1994; Evans, 2014). Today's attitude is beginning to mesh with the multicultural dictum that the only thing we share is that we differ.

The generativists began with phrase structure rules based on word classes (noun, verb, etc.) and word order. Seemingly related constructions (such as active and passive clauses in English) were related by transformations. It was soon seen that no transformation was meaningless or functionless. And the phrase structure rules were not predictive of languages that lacked phrase structure. Such languages were called "non-configurational" (Hale, 1982).

If now the quest for structural universals is being questioned, what about those rooted in meaning and function? The literature here is vast and the enterprise successful, but hardly heard of among the educated public. So let's peek in on some of the excitement. Let us consider grammatical relations (subject, object, etc.). From a structural sense, the English subject is defined as the NP (Noun Phrase) directly dominated by S, i.e., derived from the phrase structure rule S → NP + VP, with VP for Verb Phrase. It's the old rule that a sentence consists of a subject and a predicate. Is such a concept of subject universal? In light of Basque or Georgian (or Eskimo or Nez Perce or Sumerian), one has to say, no. So let's look a little deeper from the standpoint of semantics.

2   Verbal Valence

All languages have clauses that correspond to the proposition in logic and the function in mathematics. The verb names the proposition (event or state) and part of its meaning involves the number of "arguments" (participants) implied by that event or state—it's called *valence* on analogy with chemistry. We owe the concept to the French linguist Lucien Tesnière (Tesnière, 2015).

An intransitive verb (sit, exist, run, die...) implies one argument; a simple tran-

---

[1] The movement kicked off with the volume edited by Charles N. Li, *Subject and Topic* (1976); and the four volumes edited by Joseph Greenberg, Charles Ferguson, & Edith Moravcsik, *Universals of Human Language* (1978). T. Givón founded the series *Typological Studies in Language* at John Benjamins Publishing Company.

sitive verb (read, see, receive, kill...) two arguments; and a ditransitive verb (give, show, teach, tell...) three arguments. Languages do not have verbs with valence above three, which evidently has something to do with how we process information, limited chunk by limited chunk (as also to the number of primary semantic roles, see below).

The commercial event involves four parties, but we refer to it from different perspectives with verbs that imply no more than two or three of them (buy, sell, pay, cost...), and if we wish to mention all four parties in a single clause we tack on oblique prepositional phrases:

- Henry bought his wife a ring (for $1000).
- Henry paid $1000 (for a ring) (for his wife).
- They sold the ring (for $1000) (to Henry) (for his wife).

Tesnière distinguished between the arguments central to a verb's meaning (*actants*) and those peripheral to it (*circonstants*) that might be added to a clause. The number of arguments central to a verb's meaning constitutes its valence. English speaking linguists now refer Tesnière's actants as "core" arguments and his *circonstants* as "oblique" arguments. Time, place, and manner can be tacked on to just about any proposition.

- I arrived (in the morning) (at about 5:00).
- They immigrated (to another land) (with some trepidation).

A meteorological event can be construed as having a zero valence, though most languages will put the verb in the third person or—as in English—use a dummy subject *it*: "It is storming". English allows cognate objects for one place predicates (verbs of a single valence).

- He ran the race.
- She sang the song.
- He died the death of a thousand cowards.

3 Semantic Roles

Just as the variables (x, y, z) of a mathematical function bear particular relationships to the function $(+, -, / \ldots)$, the same is true of the arguments of a verb.

But in human language these relationships boil down to three primary semantic roles: Agency, Consciousness, and Neither of the above. We commonly call these Agent, Dative, and Patient (see Figure 11.1).[2] These are the relationships that exist

---

[2]The terminology varies some among linguists. For the terms here, see Talmy Givón, *Syntax* (2 volumes, 2001).

Figure 11.1: Semantic Roles

Agent	An animate instigator of an event.
Dative	An argument whose consciousness is relevant to a proposition.
Patient	Neither of the above.

between the participants in an event or state.[3] David M. Perlmutter and Paul M. Postal made these the primitives of their highly heuristic model, *Relational Grammar*, and called them 1, 3, and 2 (Perlmutter, 1983; Perlmutter and Rosen, 1984; Postal and Joseph, 1990).

The single argument of an intransitive verb might be any of these semantic roles:

- He crouched. (Agent)
- She blushed. (Dative)
- He died. (Patient)

The two arguments of a uni-transitive verb might be as follows:

- She embarrassed me. (Agent, Dative)
- He killed the goose. (Agent, Patient)
- I see a rainbow. (Dative, Patient)

The two arguments in an equative clause are Patients:

- Higgins was a doctor. (Patient, Patient)
- The book cost $20.00. (Patient, Patient)

In a ditransitive clause all three semantic roles occur: Agent, Dative, and Patient:

- She gave the check to me. (Agent, Patient, Dative)
- I will show you these old pictures. (Agent, Dative, Patient)
- He taught them linguistics. (Agent, Dative, Patient)

In an English ditransitive clause, the Dative argument can be expressed as an indirect object with preposition "to", or via "Dative-Shift" it might itself be the direct object.

- He gave some money to the landlord.
- He gave the landlord some money.

---

[3]It was the Prague linguistic circle, or Prague school, that began distinguishing semantic roles from grammatical case (nominative, accusative, etc.). The notion gained traction in the United States through Jeffery Gruber's *Studies in Lexical Relations* (1965), and Charles Fillmore, "The Case for Case" (1968).

Dative shift only works when the indirect object is conscious, i.e., a semantic Dative. The second sentence below is odd because "the top of Mt. Everest" is not conscious.

- They sent an expedition to the top of Mt. Everest.
- They sent the top of Mt. Everest an expedition.

These semantic roles are universal. How do we know? Sorry, but we just know. However they do predict grammar across multitudes of languages. Ordinary people can readily identify them. I have handed out pieces of written discourse with nouns and pronouns underlined so that students might label their semantic roles. Better than almost anything, they get it right, from the dullest to the brightest. One suspects that highly educated intellectuals would have more trouble.

4   Syntactic Primitives

Verbs with a valence of 1 will have one argument in a clause; let us label this S. Then verbs with a valence of 2 will have two arguments in a clause; let us label them A and O (Dixon, 1994). How might we distinguish A from O? They are distinguished by the following accessibility hierarchy (Keenan and Comrie, 1977):

Agent $\subset$ Dative $\subset$ Patient

The argument whose semantic role is higher on the hierarchy we label A and the one lower we label O. Thus "eat" has syntactic primitives A (Agent) and O (Patient); "see" has syntactic primitives A (Dative) and O (Patient); "insult" has syntactic primitives A (Agent) and O (Dative). Subject and object are the A and O in each of the following English clauses. They differ in their semantic roles.

- The ogre ate the pickle.
- The man saw the mountain.
- The woman insulted the stranger.

The difference between "see" and "look" is that the subject of "see" is a semantic Dative whereas the subject of "look" is an Agent. You can look but not see but you cannot see and not experience. There is one more syntactic primitive—the Dative argument in a ditransitive clause. Let us call it D. Thus "give" has syntactic primitives A, O, and D. An English ditransitive clause expressed with the least morphology— without a preposition "to"—will have a word order A + V (for verb) + D + O, with semantic roles in the same order as in the accessibility hierarchy given above.

- Bill gave Susan money.

Figure 11.2: Syntactic Primitives

A	Core argument of transitive verb highest on accessibility hierarchy.
O	Core argument of transitive verb lowest on accessibility hierarchy
S	Core argument of an intransitive verb
D	Core argument of ditransitive verb with relevant consciousness

5　Alignment

Now for the fun part. Languages differ in how they mediate between syntactic primitives and semantic roles. This is called alignment. English, for example, blurs the distinction between A and S and treats O in a special way. The pronouns 'I' and 'he' serve for both S and A, and 'him' and 'me' serve for O. The subject in English links S and A and the direct object codes for O.

I ran.　　I saw him.

He ran.　He saw me.

Basque, on the other hand, blurs the distinction between S and O and treats A in a special way. Basque *ni* corresponds to English 'I' in an intransitive clause and 'me' in a transitive clause. Basque *nik* 'I' serves for the A in a transitive clause.

- *Ni etorri naiz* 'I have come'
- *Ni ikusi zidan* 'he has seen me'
- *Nik ikusi dut* 'I have seen him'

Figure 11.3: First Person Singular Pronouns in English and Basque

	A	S	O
English	*I*		*me*
Basque	*nik*		*ni*

Let us further illustrate with examples from modern Hebrew and ancient Sumerian. In the Hebrew clauses below, S and A are treated the same (המלך *ha-mélex* 'the king') and O (את הבית *et ha-báyit* 'the house') is distinguished by the preposition *et*. In the Sumerian examples, S (𒈗 *lugal* 'king' in the first example) and O (𒂍 'house' in the second) have no case marker, whereas A (𒈗𒂊 *lugal-e* 'king' in the next example) does. The Hebrew preposition את *et-* is an accusative case marker and the Sumerian postposition 𒂊 *-e* is an ergative case marker.

Hebrew	Sumerian
המלך הלך	𒈗𒁀𒄄
ha-mélex halax	*lugal ba-ĝen*
'the king went'	'the king went'

המלך בנה את הבית	𒈗𒂊𒂍𒈬𒌦𒆕
ha-mélex bana et ha-báyit	*lugal-e é mu-un-dù*
'the king built the house'	'the king built the house'

Nez Perce is one of the very few languages that treat S, A, and O distinctly.

Figure 11.4: Nez Perce Core Cases

	'chief'	'house'
Absolute	*miyóoxat*	*iníit*
Ergative	*miyóoxatom*	*iníinm*
Accusative	*miyóoxatona*	*iníine*

- *hikúye miyóoxat* 'the chief went'
- *himéeqis híiwes iníit* 'the house is big'
- *miyóoxatom páaniya iníine* 'the chief built the house'
- *ehéxne miyóoxatona* 'I saw the chief'

Basque links S and O and treats A separately. We call the S and O case the *absolutive* and the A case the *ergative*. Hebrew, English and most European languages blur the distinction between A and S. This is called the *nominative* case. The O case is the *accusative* case.

Figure 11.5: Core Noun Cases

	A	S	O	D
Nominative	✓	✓		
Absolutive		✓	✓	
Accusative			✓	
Ergative	✓			
Absolute		✓		
Dative				✓

Some languages have "dative subject" constructions, thus the archaic *me thinks* or the German *mir ist kalt* and Russian мне холодно 'I am cold' and the Spanish *me gusta...*'I like...'. There is also an active-stative typology where intransitive subjects occur in different noun cases depending on their semantic roles. Topic marking in Philippine languages and obviation of certain Amerindian languages provide interesting case studies (Dixon, 1994; Klimov, 1974; Schachter, 1976; Morgan, 1991).

Although grammatical relations as they occur in English (subject, object, etc.) are not universal concepts, they can be defined with reference to universals of valence, semantic roles and syntactic primitives. This way grammatical relations make sense cross linguistically.[4]

Figure 11.6: Some Variations in Syntactic Alignment

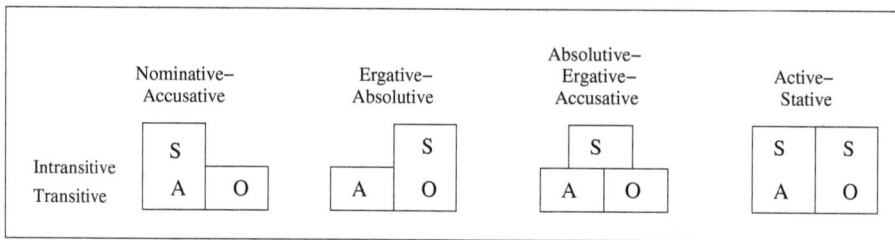

Grammar looked at this way is not a minimalist program. At a deep level it involves semantics—things that, though we can deny them, we cannot help but know them. And it involves structure. Verb valence is semantic structure that is instantiated syntactically as S, A, and O. These are structural entities, and though A and O have semantic roles higher and lower on the accessibility hierarchy, they are not themselves specifically Agent and Patient. A syntactic A is not necessarily a semantic Agent, and a syntactic O is not necessarily a semantic Patient. Alignment, along with the meaning of a verb, specifies the semantic roles in a clause.

6   Topicality

Three levels of topicality are observable across languages: primary topic, secondary topic, and non-topic. Those who sign for the deaf tell me that these are represented by three spatial positions near the signer where a referent just signed can be placed and recovered later. I will not elaborate here except to say that voicing mechanisms (active, passive, antipassive, direct, inverse, applicative, directive...), definiteness marking, and varying other modes of topicalization are all involved in third person referent tracking (i.e., in the way we keep track of what we are talking about). Grammars mediate between semantic case roles and topicality, or else we wouldn't

---

[4]The Wikipedia article is pretty good: https://en.wikipedia.org/wiki/Ergative

know cause from effect or maintain coherence (which maybe I'm not maintaining very well right now, come to think of it).

7 Coding

Grammar codes in two main areas:

1. Grammatical relations which mediate between semantic roles (agency, consciousness) and principles of coherence;

2. Tense-aspect-modality (TAM), which is concerned with time, sequencing, possibility, predictability, necessity, obligation, etc.

Language is a code. And like all codes, language codes information. Spoken language exploits three aspects of the stream of sound:

1. Word order

2. Affixation & periphrasis (prefixes, suffixes, prepositions, auxiliaries...)

3. Prosody (intonation, nasalization, creaky voice, breathy voice...)

If reality is "it from bit" (Wheeler, 1990), language is the latter, though it mostly codes for something else. The major thing that language codes, and that which is thus imprinted on the grammars of all natural languages, is mind. What people are most concerned with is agency and consciousness. Mathematics is human language formalized but minus these two.

The quest for universals has not been a failure. There is much that all languages share, this despite what we hear so often today. People are not robots. And so there are always unpredictable features of the language they use—in the code they have inherited and in their personal usage—thus Edward Sapir's celebrated statement:

> Were a language ever completely 'grammatical' it would be a perfect engine of conceptual expression. Unfortunately, or luckily, no language is tyrannically consistent. All grammars leak.

> (Sapir, 1921)

There is nothing new that I have said here; the literature, as I said earlier, is vast. Nevertheless, it seemed good to explain this to the educated non-linguist who knows none of this and might be tempted to doubt that he is conscious and a free moral agent.

References

Bloomfield, L. 1933. *Language.* Henry Holt, New York.

Chomsky, N. 1959. A review of B. F. Skinner's *Verbal Behavior. Language* 35(1):26–58.

Croft, W. and Cruse, D.A. 2004. *Cognitive Linguistics.* Cambridge University Press.

Dixon, R.M. 1994. *Ergativity*, volume 69 of *Cambridge Studies in Linguistics.* Cambridge University Press.

Evans, V. 2014. *The Language Myth: Why Language is Not an Instinct.* Cambridge University Press.

Evans, V. and Green, M. 2006. *Cognitive Linguistics: An Introduction.* Edinburgh University Press, Edinburgh.

Fillmore, C. 1968. The case for case. In E. Bach and R.T. Harms (editors), *Universals in Linguistic Theory*, pp. 1–88, Holt, Rinehart and Winston, New York.

Givón, T. 2001. *Syntax.* John Benjamins Publishing Company, Amsterdam, revised edition.

Greenberg, J., Ferguson, C., and Moravcsik, E. 1978. *Universals of Human Language.* Stanford University Press.

Gruber, J. 1965. *Studies in Lexical Relations.* MIT, ph.D. Dissertation.

Hale, K. 1982. Preliminary remarks on configurationality. *Proceedings of the North Eastern Linguistic Society* 12:86–96.

Harris, R.A. 1993. *The Linguistic Wars.* Oxford University Press.

Keenan, E.L. and Comrie, B. 1977. Noun phrase accessibility and universal grammar. *Linguistic Inquiry* 8(1):63–79.

Klimov, G.A. 1974. On the character of languages of active typology. *Linguistics* 131:11–25.

Lakoff, G. and Johnson, M. 1999. *Philosophy in the Flesh: the Embodied Mind and its Challenge to Western Thought.* Basic Books, New York.

Levine, R.D. and Postal, P.M. 2004. *A Corrupted Linguistics.* Corrupted Linguistics, San Francisco.

Li, C.N. 1976. *Subject and Topic.* Academic Press, New York.

Morgan, L. 1991. *A Description of the Kutenai Language.* University of California, Berkeley, phD dissertation.

Newmeyer, F.J. 1996. *Generative Linguistics: A Historical Perspective.* Routledge, London.

Perlmutter, D.M. 1983. *Studies in Relational Grammar 1.* University of Chicago Press.

Perlmutter, D.M. and Rosen, C.G. 1984. *Studies in Relational Grammar 2.* University of Chicago Press.

Pinker, S. 1994. *The Language Instinct: How the Mind Creates Language.* William Morrow and Company, New York.

Postal, P.M. and Joseph, B.D. 1990. *Studies in Relational Grammar 3.* University of Chicago Press.

Sapir, E. 1921. *Language: An introduction to the Study of Speech.* Harcourt, Brace and Company, New York.

Schachter, P. 1976. *The subject in Philippine languages: Actor, topic, actor-topic, or none of the above.* Academic Press, New York.

Tesnière, L. 2015. *Elements of Structural Syntax.* J. Benjamins.

Wheeler, J.A. 1990. *Information, Physics, Quantum: the Search for Links.* Addison-Wesley.

12

Psychology: Methodological Dualism and a Multi-Explanation Framework—An approach needed for understanding behavior

SAM S. RAKOVER

Haifa University

Abstract

At present there is no explanation for the Mind/Brain relationship. It is hard to conceive of mentalistic explanations in terms of mechanistic explanations, where mechanistic explanations refer to explanations common in the sciences such as neurophysiological and computational, and mentalistic explanations are based on the individual's inner world such as will, belief, intention, and purpose. And, it is difficult to provide a comprehensive explanation of behavior and its components by an appeal to mechanistic explanations only. Therefore, it makes sense to develop a new methodological approach, *Methodological Dualism*, which leads to the construction of a *Multi-Explanation Framework* for developing specific psychological theories. This approach is not based on the usual attempt to reduce mental processes to neurophysiological processes. It addresses complex behavior by means of multiple explanations (mechanistic and mentalistic), which are appropriately matched to behavior and its components. This approach offers a deeper understanding of behavior than that provided by mechanistic explanations alone.

1 Introduction

The present paper is based on my previous publications and several further developments (Rakover, 2007, 2011/2012a, 2011/2012b). It consists of four parts. The first part presents the rationale for developing Methodological Dualism (MD) and the Multi-Explanation Framework (MEF). It is based on a sort of argument, called the "guiding-argument," which suggests reasons and justifications for developing MD and MEF. The second and third parts present the major ideas of MD and MEF, respectively. Finally, the fourth part discusses the major contributions of this approach to psychology and responds to criticisms.

2 Rationale for Developing MD and MEF

The rationale is based on the following guiding-argument that leads to several conclusions:

Assumptions

A. To date there is no accepted theory that explains the relationship between mind and body, consciousness and brain [there is no T(m/b)].

B. It is hard (if not impossible) to conceive of mentalistic explanations (e.g., based on our own will, belief, intention, consciousness) in terms of mechanistic explanations (e.g., based on events or processes that are physical, neurophysiological, cognitive-computational).

C. Behavior is saturated with consciousness experiences. We are aware of our everyday behavior, the environment in which we live, the relations between us and the environment (physical, social), and in many cases we are conscious of our own consciousness. We act to change our consciousness and we are driven by it.

Conclusions

1. If one suggests only a mechanistic explanation for behavior (in the manner accepted by Behaviorism, Cognitive Psychology and Physiological-Psychology), the explanation will be incomplete because it disregards consciousness.

2. If one aspires to present a better (approaching complete) account of behavior, one has to take into consideration consciousness, i.e., to offer mentalistic explanations in addition to mechanistic accounts.

3. Since, to date, there is no T(m/b) and a mentalistic explanation cannot be based on a mechanistic one, a special model of a mentalistic explanation has to be developed if one aspires to offer a better account of behavior.

4. In view of the above, MD and MEF have to be developed.

The assumptions of the guiding-argument are supported by many professional publications, which are briefly summarized here. We are still incapable of understanding the mind by means of the brain and consciousness by means of neurophysiological-computational processes occurring in the brain. To date, no one has succeeded in explaining conscious mental states and processes by physical and neurophysiological concepts (Bayne, 2009; Palmer, 1999; Rakover, 1990, 1997, 2007). We still have no T(m/b) detailing the mechanism whereby neurophysiological activity in the brain creates or acquires consciousness akin to the accepted theories in science: physical theories that explain the transformations of energy (associated with potential and kinetic energy, friction and heat, magnetism and electricity, mass and energy) or how matter changes from one kind to another, like the theory of how hydrogen and oxygen join to form water, and how water can be broken down (by electrolysis) into these gases. Utall (2014) proposed a negative answer to the question in his paper, titled: "Are neuroreductionist explanations of cognition possible?" for methodological, conceptual, and empirical reasons. However, I do agree with McCauley and Bechtel (2001) that if, indeed, it were possible to reduce a psychological theory to a neurophysiological theory one could forgo psychological concepts altogether since behavior would be explained through the theories prevailing in the sciences. But this is not how matters stand at present.

Several researchers have suggested that wide-ranging activity of a neural network that unites various functions in the brain (such as the "global neural workspace") is a source of consciousness (Baars, 1988, 2002; Cosmelli, Lachaux, and Thompson, 2007; Dehaene and Naccache, 2001; Kouider, 2009; McGovern and Baars, 2007; Palmer, 1999). However, this proposal is based on a correlation between neurophysiological indexes (e.g., neuroimaging) and expressions of consciousness, and is not a T(m/b) (Cosmelli et al., 2007; Miller, 2011). Similarly, research in cognitive modeling or artificial intelligence has not succeeded in solving the consciousness problem (McDermott, 2007; Searle, 1980, 1990; Sun and Franklin, 2007). One may suggest that it is possible to conceive mentalistic explanations in terms of mechanistic explanations. Consider, for example, the following purposive (teleological) explanation: David drove to Tel Aviv in order to meet Ruth. This purposive explanation can be equivalently expressed (translated) as the causal explanation: the thought of meeting Ruth in Tel Aviv caused David to drive there. However, the translation is problematic for at least the following two reasons: First, it is generally assumed that similar causes produce similar effects. Nevertheless, this assumption does not hold here. While the goal of meeting Ruth can be achieved in many different ways, a cause—which is the translation of a goal—does not produce such a diversity of effects. Secondly, while it is generally assumed that the cause and effect are separate and different events, this assumption is not maintained here. David's will, belief, and action are interconnected. The drive to Tel Aviv is described as a meaningful action that involves David's will and belief, and similarly, a meaningful description of each of these three terms engages the other two. (For a discussion of this and related issues see Rakover, 1990, 1997, 2007.) Furthermore, it has been found that although mental processes are involved in behavior, attempts to explain mental causation (i.e., how mental properties causally affect motor and physiological behavior) encountered impassable obstacles. There are well-known problems that show that mental causation is impossible to account for, such as the problem of "anomalism" that is based on the absence of strict psychophysical laws, etc. (For review see Robb and Heil, 2014; Yoo, 2015.) Thus, on the basis of this brief review it is safe to accept the above assumptions of the guiding-argument and propose that the stage is set for the development of Methodological Dualism.

3   Methodological Dualism

3.1   Meeting the Methodological Requirements for Scientific Explanation

The major goal of Methodological Dualism is to develop three cornerstone ideas for establishing the proposal that a mentalistic explanation (a purposive explanation) does meet the methodological requirements for scientific explanation.

3.1.1   Making a distinction between a specific explanation and an explanation model (procedure)

A model of explanation is a general procedure for creating different specific explanations for different specific empirical observations. Only explanation procedures (models, schemes) that fulfill the methodological requirements of science are approved and accepted by the scientific community. Here I concentrate on two different models (schemes, procedures) of explanation: the D-N model and the purposive (will/belief) scheme. The first model, widespread in the natural sciences, is the Deductive-Nomological (D-N) model (see Hempel, 1965). It proposes that a specific explanation (i.e., prediction, called the dependent variable in psychology) is deduced from a law (theory) together with particular conditions (i.e., independent variables).

The second procedure (scheme, model) creates specific purposive will/belief explanations, which are widespread in folk psychology. For example, David drove to Tel Aviv because he wanted to meet Ruth and believed that a drive there would realize his will. Here, I propose to conceive as a mentalistic explanation-model the procedure for creating specific will/belief explanations:

[Will/Belief] If X wants G and believes that behavior B will realize his/her will, then X will perform B.

This is a new proposition, and it is central to the present approach. It immediately raises the following question: Does this explanation procedure satisfy the methodological requirements for explanation accepted in science?

The professional literature can be interpreted as suggesting a negative answer because it takes [Will/Belief] as a social-scientific law that can be placed in the D-N model. But as we shall see below, the answer is affirmative: I argue that [Will/Belief] does satisfy the accepted methodological requirements of a scientific explanation model and not of a social-scientific law.

3.1.2 Making a distinction between mechanistic and mentalistic explanation procedures

As mentioned above, mechanistic explanations, which are common in the natural and social sciences, can offer explanations for behavior of humans and animals by appeal to physical, chemical, physiological, genetic, evolutionary factors, as well as to stimulus-response-consequence relations, which reject the use of mental terms in explanations, and to cognitive-computational processes analogous to the workings of a computer, such as symbolic (classic) models or neural networks (see Bechtel, 2008; Rakover, 2007, 2011/2012a, 2011/2012b). Mentalistic explanations offer everyday accounts for people's behavior by appeal to their mental conscious states and processes, such as will/belief. For example: David drove to Tel Aviv in order to meet Ruth. In this case, the public behavior, "David drove to Tel Aviv," is explained by an appeal to David's conscious experience: David's will to meet Ruth and his belief that the trip to Tel Aviv would realize his will. Underlying this and other examples is the assumption that the individual is endowed with mental conscious states and processes (will, belief, purpose, intention, thought, emotion, etc.) that are the basis of a mentalistic explanation.

3.1.3 Developing justifications for conceiving the mentalistic scheme [Will/Belief] scientifically

It is revealed that [Will/Belief] fulfills the requirements of the scientific methodology for providing explanations. Although it is very difficult to reduce a mentalistic explanation to a mechanistic explanation, it is discovered here that the scientific requirements for explanation are wide enough to encompass the mentalistic explanation scheme [Will/Belief] too. Hence, according to the scientific game-rules, mentalistic explanations are methodologically suitable. In the following, I present arguments that support the proposal that [Will/Belief] (a) meets the methodological requirements for a procedure (model) of scientific explanation, and (b) cannot be conceived of as similar to a law in the sciences.

3.2   Methodological requirements for a procedure of scientific explanation

Based on the literature on explanation, I propose that an explanation procedure (scheme) has four major characteristics (Hempel, 1965; Lipton, 1992, 2001; Psillos, 2002; Rakover, 1990, 1997; Salmon, 1989; van Fraasen, 1980; Woodward, 2002). I add a fifth requirement called "Empirical Irrelevance."

General procedure An explanation model (scheme) is a general procedure whereby the researcher proposes specific explanations for specific phenomena. This property exists also in [Will/Belief]. For example: David wants to bid farewell to Ruth so he waves his hand.

Causes and reasons An explanation model in the natural sciences assumes that in most cases the explanation for a phenomenon is associated with a general law, a theory, or a mechanism, which proposes a cause for the phenomenon's occurrence. Analogously, regarding a human's or an animal's behavior, the explanation is accomplished by an appeal to internal mental processes that explain it or give reasons for it.

Rationality In the natural sciences an explanation procedure, scheme, or model creates from one sort of information (the *explanans*) another sort of information (the *explanandum*) by means of rules of logic (deduction, induction), mathematics, and probability. Thus, the occurrence of the studied phenomenon is expected because it is predicted on the basis of specific information and rational rules. Similar things happen in the case of the teleological explanation. However, the prediction based on [Will/Belief] is not based on logic, on statistical probability, or on the necessity that derives from a natural law. Rather, it is based on practical reasoning, on the considered opinion of the individual who takes into account, among other things, his ability, the physical and social conditions to which he is subject, and the significance of realizing his will (see Millgram, 2001; Newell, 1981; Samuels, Stich, and Faucher, 2004; Schueler, 2003; von Wright, 1971).

Empiricism The specific explanation generated by the explanation procedure (scheme, model) must be attached to reality. This enables an empirical test of the theory, the law, the mechanism that the explanation model uses. This requirement is also realized for the mentalistic explanation, [Will/Belief]. For example, there is no problem in testing empirically the explanation that David waved his hand as a sign of his wishing to take leave of Ruth.

Empirical Irrelevance To use the D-N model, one has to set in the model's assumptions various laws and theories (e.g., laws of the movement of bodies, laws in electricity or electromagnetism, theories or laws in biology, etc.) and the relevant particular conditions, and then derive from them specific predictions. This model, then, possesses the property of being an explanation storehouse for diverse theories and laws from many and varied fields of research. Similarly, one may suggest that a hypothesis (theory or law) is confirmed or refuted by the familiar Hypothetico-Deductive (H-D) method. This method also owns the property of being a storehouse for empirical testing of diverse hypotheses, theories, and laws. These properties underlie the present characteristic, Empirical Irrelevance.

These properties suggest that observations do not relate empirically to the explanation model and the method of testing themselves, but only the hypothesis, the theory, or the law, inserted in these procedures. That is, methodologically the empirical observations are not relevant to the explanation procedure (scheme) and the method of testing; they do not qualify/disqualify the procedures for explanation or testing. If observations were relevant to the hypothesis (theory, law), the explanation procedure, and the method of testing, then one discordant observation (i.e., negative empirical result) would refute, in addition to the hypothesis, the explanation procedure and the method of testing. Therefore, no hypothesis (theory, law) could then be tested and used for explanation since the testing and explanation procedures would be eliminated.

Does the Empirical Irrelevance attribute also apply to the procedure of explanation: [Will/Belief]? In my opinion, the answer is affirmative. Consider the following example: David wants to meet Ruth in Tel Aviv and believes that a bus ride will realize his wish. Hence, a specific prediction may be proposed that David will travel to Tel Aviv. But, David does not travel to Tel Aviv. According to the present attribute, what was refuted was the specific hypothesis about David's travel and not the scientific method of testing or the teleological explanation scheme, [Will/Belief], whereby this specific prediction was generated. The reason for this is similar to what was stated above: the teleological explanation scheme continues to produce specific teleological explanations, specific predictions, which deal with other behaviors of David (and of other people). Otherwise, it would not be possible to put any specific teleological hypothesis to an empirical test, because in principle one negative result is sufficient to refute the specific hypothesis, the method of testing, and the teleological explanation scheme, [Will/Belief], that generated the specific explanation.

3.3 [Will/Belief] cannot be conceived of as similar to a law

The question that arises here is whether the present interpretation of [Will/Belief] as an explanatory procedure is the only one. Several researchers have formulated a purposive explanation, a [Will/Belief] explanation, in a way similar to a law in the natural sciences (e.g., (Churchland, 1988; Horgan and Woodward, 1985; Rosenberg, 1988)). I cannot accept this interpretation for the following reasons.

First, if the Empirical Irrelevance characteristic holds, then [Will/Belief], conceived of as an explanation procedure (model, scheme), is not empirically testable, whereas all laws, theories, and hypotheses are empirically testable. Therefore, [Will/-Belief] may not be conceived of as a kind of scientific law. Contrary to Churchland (1988), who maintains that folk psychology is unchanging because it is fundamentally bad science, and its fate is to disappear from the book of science just as popular theories about ghosts disappeared, I argue, in accordance with Empirical Irrelevance, that [Will/Belief], as an important part of folk theory, is irrefutable not because folk theory is bad science, but because [Will/Belief] is a mentalistic procedure for generating various specific explanations and is not affected by empirical results.

Secondly, [Will/Belief] does not seem to uphold the criteria of scientific laws (see Swartz, 1985; Weinert, 1995; Woodward, 2000, 2003). In addition to these, I offer the criterion of "unit equivalency" (see Rakover, 2002) as another reason why is it difficult to conceive of [Will/Belief] as a law.

Physical laws and theories uphold the requirement that I call "unit equivalency" (see Rakover, 1997, 2002), whereas psychological theories do not. (The present requirement corresponds to Dimensional Analysis used in the sciences for checking the correctness of equations.) Unit equivalency requires that the combination of measurement units on one side of the equation of the law, or theory, must be identical to the combination of measurement units on the other side. To clarify this requirement, let us consider Galileo's famous law of free fall of bodies: $d = \frac{1}{2}gt^2$, where d signifies distance, t time, and g acceleration of the body caused by gravitation.

Since d is measured in units of distance (meters), the expression gt^2 must also be measured in units of distance. And indeed, a simple algebraic calculation shows that this is the case:

$$\text{meter} = \frac{\text{meter}}{\text{time}^2} \cdot \text{time}^2$$

Does [Will/Belief] satisfy this requirement? Is the combination of measurement units on one side of the equation Action = f(Will, Belief) equivalent to the combination of measurement units on the other side of the equation? The answer is negative. The combination of measurement units common in psychology for concepts of will and belief (usually measured by verbal report) is not identical to the combination of the measurement units of the term action (usually measured by frequency of response, reaction time, etc.). Nevertheless, one may suggest certain interpretations for the coefficients of the equation Action = f(Will, Belief) (coefficients that multiply the expression including will, belief) so that the present requirement for unit equivalency can be satisfied. However, this possibility is no more than *ad hoc*, since in psychology the interpretations and estimations of coefficients (as in regression) change from situation to situation. This in no way parallels the interpretations and estimations of the coefficients common in the sciences, which are invariant and universal.

Thus, good justifications have been presented for treating [Will/Belief] as a procedure for scientific explanation: it maintains the methodological properties of an explanation model accepted in science and not of a scientific law. This sets the stage for the development of the Multi-Explanation Framework discussed in the next section.

4 Multi-Explanation Framework

The present section develops MEF, which is founded on the following four major ideas.

4.1 Using the MEF as a Framework for Theory Development

The MEF is not a specific theory in a specific psychological area, but is a framework that allows one to develop a specific theory in a specific area of psychology. It suggests guidelines and procedures on how to construct coherently a specific theory based on two kinds of explanations, mechanistic and mentalistic, and how to test it empirically. Although most current psychological theories are based on mechanistic explanations, nearly all behaviors need to be accounted for by appeal to both mechanistic and mentalistic explanations. The MEF is based on a match between explanation procedures (schemes, models) that can be mechanistic or mentalistic and behavior. Let us call it the "explanation/behavior match" for short. While there are many behaviors that can be accounted for satisfactorily by appeal to mechanistic explanations only (several such explanatory models are known in the literature, e.g., Salmon, 1989), many behaviors and their components need to be approached by both mechanistic and mentalistic explanations. The coherence of a specific theory based on the MEF is achieved by matching appropriately the explanation procedure to the behavior and its components by means of several guidelines and procedures, to be detailed below.

4.2 Employing Multiple Explanation Procedures

In the natural sciences an explanation model or procedure employs various laws or theories to suggest mechanistic explanations for various specific phenomena. By contrast, the MEF posits that behavior has to be understood by appeal to a theory that employs coherently several explanation procedures—mechanistic and mentalistic. While the natural sciences and the present framework differ methodologically in providing explanation, they do not differ in the methodology of empirically testing a theory.

A psychological theory based on the MEF can be tested in a way similar to testing a theory in the natural sciences. The testing method (the Hypothetico-Deductive

[H-D] method) is indifferent to the kind of theory the researcher uses to explain the experimental results. The H-D method can be used as long as a prediction can be derived from the theory under study and the prediction can be compared with the observation. For example, there is no problem testing empirically the [Will/Belief] explanation that David waved his hand as a sign of his wishing to take leave of Ruth. Since David is acquainted with Ruth, one can predict that he will recognize (choose correctly) her photo out of 10 different photos. And Ruth, who will recognize David's photo as well, will confirm that she saw David waving his hand to say goodbye by waving back.

However, a complete explanation of David's behavior is different from the unified explanation given to a flashlight, for example. The explanation for the latter is provided by decomposition of the flashlight into parts (battery, bulb, etc.) by providing an explanation for the operation of each part through the appropriate mechanistic theory and by giving an explanation for the interaction between these parts through calculation of transformation of energy. This procedure does not work with a MEF explanation of David's behavior because we still do not know how mental processes (e.g., David's will or belief) interact with neurophysiological processes (e.g., David waved his hand) and how consciousness emerges from the brain. In short, we still do not possess a T(m/b), and therefore we cannot provide a complete and unified explanation of behavior as we could with a flashlight.

4.3 Matching Explanations to Components

As mentioned above, the MEF is based on a match between explanation procedures (mechanistic or mentalistic) and behavior—the explanation/behavior match. To accomplish this match, one has to break a behavior down into its components and fit the appropriate mechanistic or mentalistic explanations to the whole behavior and its components. This involves several steps: (1) One determines whether the studied behavior matches mechanistic or mentalistic explanation procedures (schemes, models). (2) One chooses from several explanation schemes the particular one that fits the studied behavior and its components. I call the matching pair explanation/behavior an "explanation-unit." (3) Finally, one organizes coherently the explanation-units in the specific theory developed for the studied behavior.

How then does one perform these steps and fit explanations to behavior and its components? Because the answer depends on many factors (associated with theoretical and empirical knowledge), a general solution to this issue cannot be proposed. Accordingly, I shall describe, in brief, two important indicators suggesting when the explanation of a given behavior will be achieved with the aid of a mechanistic explanation scheme and when with the aid of a mentalistic scheme.

Indicator by empirical research Numerous studies discuss the following questions. Is one capable of being aware of the five events connected to one's own behavior: the presented information (the stimuli: verbal, visual, etc.), the response to this information, the mechanism that produces the response, the purpose of the response, and the result caused by the response. Can one control (initiate, change, stop) these five events? The answers to these questions may guide the researcher in the choice of the proper explanation (mechanistic, mentalistic). Here are several examples illustrating the above five events. The examples start with behavior that is explained mechanistically and end with behavior that is accounted for mentalistically.

In many cases the individual is not aware of, and does not control, the five foregoing events, or the larger part of them. These events are associated with chemical, neurophysiological processes in our brain and body and are accounted for mechanistically (see Morsella, 2005; Palmer, 1999; Rakover, 1983).

A number of cases are characterized by cognitive impenetrability (Pylyshyn, 1984). If a given behavior is not affected by a change in the individual's goals, beliefs, desires, thoughts, and knowledge, it is reasonable to assume that it is based on innate mechanistic processes (see also Fodor, 1983).

In many cases one plans a behavior that will satisfy one's conscious [Will/Belief]. In fact, many of one's actions (e.g., reading, listening to music, watching plays and movies, travel and touring, meeting friends, etc.) are performed to change one's own conscious experience.

Indicator by the principle of explanation-matching This principle deals with cases where behavior is divided into its components, each of which has to be assigned the appropriate (mechanistic or mentalistic) explanation. The following question then arises: What is the relation between the kind of explanation that has been matched to a whole behavior (A) and the kinds of explanation that have been matched to its behavioral components (a_1 a_2 a_3 etc.)? To resolve this problem, the principle of explanation-matching is proposed:

Mentalistic behavior Behavioral components of a mentalistic behavior—a whole behavior that was explained mentalistically—are likely to receive both mentalistic and mechanistic explanations.

Mechanistic behavior Behavioral components of mechanistic behavior—a whole behavior that was explained mechanistically—will receive only mechanistic explanations.

Accordingly, this principle posits that the components of a mechanistic behavior cannot receive mentalist explanations, while the components of a mentalistic behavior may receive mentalistic as well as mechanistic explanations. Hence, if mechanistic behavior A is broken down into two behavioral components a_1 and a_2, where a mechanistic explanation is matched to a_1 and a mentalistic explanation is matched to a_2, then either behavior A was not purely mechanistic or the match of the explanation to the behavioral component a_2 failed.

For example, since the most popular Müller-Lyer illusion is not affected by mentalistic factors such as knowledge of the structure of the illusion, and since it appears in fish and chicks too (see Coren and Girux, 1978), the illusion's explanation has to be mechanistic. Therefore, it is hard to see how the components of this illusion are likely to receive a mentalistic explanation. However, if a certain plan of action is put into practice (e.g., going to the movies) a complex network of systems (mentalistic and mechanistic) is activated that enable going to the movies and all that it involves. In this case, the whole behavior is explained mentalistically, and its components are explained mechanistically and mentalistically.

4.4　Potential Problems Arising in Using the MEF

The idea that the MEF is based on several explanation schemes (various mechanistic and mentalistic schemes) is liable to give rise to the following three problems, for which a solution is offered by the procedure of fitting the appropriate explanation scheme to behavior:

The *ad hoc* explanation problem Because the MEF contains a number of explanation models (procedures, schemes) the following problem can be raised. If a certain phenomenon is not explainable by scheme (A) one may try scheme (B), and so on until a satisfactory explanation is found. The problem created by this possibility is that the theory may propose *ad hoc* explanations and not be empirically refutable.

The inconsistency problem Since the MEF contains a number of explanation schemes; it may provide an explanation for a certain phenomenon through explanation model (a), which may stand in contradiction to explanation model (b). The theory based on this approach may thus be beset by internal contradictions.

The incomparability problem Because the MEF contains a number of explanation schemes, two specific theories that are based on this framework may not be comparable as they may employ different explanation schemes (procedures). As a result, it may be difficult to decide whether dissimilar predictions are generated by the different content of the theories themselves or by the different explanatory schemes they employed.

The main idea for solving these problems rests on the commitment to match an explanation to a behavioral component, that is, to preserve the explanation-unit. The obligation to the explanation/behavior match solves these three problems, because for every behavioral component the researcher uses only one single explanation. Note that a commitment to the explanation-unit is similar to a commitment to a hypothesis: one holds it till it is disconfirmed. As a result, the researcher can neither propose different explanations for the behavior under study, nor can she leap from explanation to explanation at will. He/she must use one explanation—the one determined as most suited to handling the kind of the studied phenomena. This commitment prevents the possibility of proposing *ad hoc* explanations. The explanation-unit is fixed in advance, as in the methodology applied in the natural sciences; there, for example, it is clear that the movement of bodies is treated by means of an explanation model of the kind proposed by Hempel (1965).

The explanation/behavior match protects the theory from the problem of inconsistency. Because for every phenomenon a matching explanation is determined in advance, no situation will arise where the researcher uses different explanations for the same behavior, hence a situation of self-contradiction will not arise.

The explanation/behavior match also allows empirical and theoretical comparisons of different theories. This comparison may be partial or entire when the two specific theories based on the MEF use the same explanation procedures (schemes).

5 Discussion

The major contribution of the present paper is in proposing a unique approach, MD, which leads to the development of the MEF. The latter helps a researcher to develop a specific psychological theory with an improved explanatory ability. The approach is not founded on the usual attempt to reduce mental states to neurophysiological states; it circumvents the ontological Mind/Body problem and the debate on dualism vs. monism. The present approach is founded on a nontrivial integration of two kinds of explanation, mechanistic and mentalistic, which enhances the theory's ability to provide a better understanding of behavior than by the standard mechanistic theories developed by behaviorism and cognitive psychology. The present approach is founded on the following ideas:

1. Various specific purposive explanations are generated by the mentalistic scheme [Will/Belief]. Although [Will/Belief] cannot be reduced to a mechanistic explanation, it does fulfill the scientific methodological requirements for explanation.

2. Given (1), one may construct the MEF, which provides guidelines on how to develop a specific psychological theory based on two kinds of explanatory models: mechanistic and mentalistic. This is achieved by making an appropriate match between behavior and the two kinds of explanations, mechanistic and mentalistic—the explanation/behavior match.

3. The explanation/behavior match establishes the coherence of the specific MEF's theory, and also offers solutions to certain problems that are raised by the proposed framework, which is founded on the multiplicity of explanation models.

Since the present approach is based on the strong empirical/theoretical impression that a mentalistic explanation is essential for understanding behavior, I will end the present paper with the severe criticism that mental explanation is otiose.

Several years ago I had a discussion with colleagues in the department of psychology at Haifa University on the question of whether conscious experiences, mental states, are needed for an explanation of behavior. They maintained that any behavior can be explained mechanistically without resort to consciousness. I did not agree. I believe that my colleagues were expressing the common approach to this issue (held by behaviorism and cognitive psychology). Why? Basically, because of the following "otiose" argument: Given that many mathematical functions can be fitted to any set of empirical observations, and given that these functions express different mechanistic theories, it follows that it is possible to construct a mechanistic theory for any set of psychological observations. Hence, one may ask why we should appeal to the complex hybrid MEF if we can explain a given phenomenon successfully by appeal to a mechanistic explanation. That is, consciousness apparently does not enhance our understanding (predicting, explaining) over and above the understanding that one obtains by a mechanistic theory. In this regard Dawkins (1995) wrote: "There is no prediction we can make that if the animal has consciousness it should do X but not conscious it should do Y" (pg. 139). Flanagan (1992) conceived of "conscious inessentialism" as "the view that for any activity i performed in any cognitive domain d, even if we do i consciously, i can, in principle, be done nonconsciously" (pg. 129). Similarly, the popular solution to the problem of mental causation by identifying a particular mental event with a particular physical event (called the token identity theory) raises the following counter argument: Given the token identity theory, one does not need the mental event to causally explain behavior simply because the complete explanatory job is done by the physical event (see Robb and Heil, 2014; Yoo, 2015). I don't agree. Mental events and processes are essential for the explanation of behavior.

First, as I illustrated above and in Rakover (2007, 2011/2012a, 2011/2012b) there are various behaviors that do necessitate an appeal to mentalistic explanations because it seems difficult to propose purely mechanistic explanations for them.

Secondly, I believe that the otiose argument is inadequate. In my view, this argument is based on a crucial hidden assumption that many researchers have overlooked: The behavior that is explained in psychology is behavior stripped of conscious meaning. The psychological indexes (e.g., number of correct responses and reaction time) do not carry any conscious meaning since these indexes are the same as those that represent a robot's behavior. They are public responses. Psychologists treat only those behavioral properties that belong to the public domain, not properties that carry private conscious meaning. Thus, if the behavior to be explained is devoid of conscious meaning, then it is no wonder that the explanation is constructed mechanistically. But, in this case, one does not account for a conscious meaningful behavior. (One may say that the explanation is given to a zombie's behavior).

References

Baars, B.J. 1988. *A cognitive theory of consciousness.* Cambridge University Press, New York.

Baars, B.J. 2002. The conscious access hypothesis: Origins and recent evidence. *TRENDS in Cognitive Sciences* 6:41–52.

Bayne, T. 2009. Consciousness. In P. Robbins and M. Aydede (editors), *The Cambridge handbook of situated cognition,* Cambridge University Press, Cambridge.

Bechtel, W. 2008. *Explanation: Mechanism, modularity, and situated cognition.* Cambridge University Press.

Churchland, P.M. 1988. *Matter and consciousness.* MIT Press, Cambridge, MA, revised edition.

Coren, S. and Girux, J.S. 1978. *Seeing is deceiving: The psychology of visual illusions.* LEA, Hillsdale, NJ.

Cosmelli, D., Lachaux, J.P., and Thompson, E. 2007. Neurodynamical approaches to consciousness. In P. Zelazo and M.M..E. Thompson (editors), *The Cambridge handbook of consciousness,* Cambridge University Press, Cambridge and New York.

Dawkins, M.S. 1995. *Unraveling animal behavior.* Essex: Longman Scientific and Technical, 2nd edition.

Dehaene, S. and Naccache, L. 2001. Towards a cognitive neuroscience of consciousness: Basic evidence and a workspace framework. *Cognition* 79:1–37.

Flanagan, O. 1992. *Consciousness reconsidered.* Cambridge, MA: The MIT Press.

Fodor, J. 1983. *The modularity of mind.* Cambridge, MA: MIT Press.

Hempel, C.G. 1965. *Aspects of scientific explanation and other essays in the philosophy of science.* The Free Press, New York.

Horgan, T. and Woodward, J. 1985. Folk psychology is here to stay. *The Philosophy Review* 94:197–226.

Kouider, S. 2009. *Neurobiological theories of consciousness.* Elsevier Press.

Lipton, P. 1992. The seductive-nomological model. *Studies in History and Philosophy of Science* 23:691–698.

Lipton, P. 2001. *What good is an explanation?* Kluwer Academic Publishers, The Netherlands.

McCauley, R.N. and Bechtel, W. 2001. Explanatory pluralism and heuristic identity theory. *Theory and Psychology* 11:736–760.

McDermott, D. 2007. Artificial intelligence and consciousness. In P. Zelazo, M. Moscovitch, and E. Thompson (editors), *The Cambridge handbook of consciousness*, Cambridge University Press, Cambridge and New York.

McGovern, K. and Baars, B.J. 2007. *Cognitive theories of consciousness.* The Cambridge handbook of consciousness.

Miller, G.A. 2011. Mistreating psychology in the decades of the brain. *Perspectives on Psychological Science* 5:716–743.

Millgram, E. 2001. Practical reasoning: The current state of play. In E. Millgram (editor), *Varieties of practical reasoning*, MIT Press, Cambridge, MA.

Morsella, E. 2005. The function of phenomenal states: Supramodular interaction theory. *Psychological Review* 112:1000–1021.

Newell, A. 1981. The knowledge level. *AI Magazine* 2:1–33.

Palmer, S.E. 1999. *Vision science: Photons to phenomenology.* MIT Press, Cambridge, MA.

Psillos, S. 2002. *Causation and explanation.* Acumen, Chesham, UK.

Pylyshyn, Z.W. 1984. *Computation and cognition.* MIT Press, Cambridge, MA.

Rakover, S.S. 1983. Hypothesizing from introspections: A model for the role of mental entities in psychological explanation. *Journal for the Theory of Social Behavior* 13:211–230.

Rakover, S.S. 1990. *Metapsychology: Missing links in behavior, mind and science.* New York: Paragon/Solomon.

Rakover, S.S. 1997. Can psychology provide a coherent account of human behavior? a proposed multiexplanation-model theory. *Behavior and Philosophy* 25:43–76.

Rakover, S.S. 2002. Scientific rules of the game and the mind/body: A critique based on the theory of measurement. *Journal of Consciousness Studies* 9:52–58.

Rakover, S.S. 2007. *To understand a cat: Methodology and philosophy.* John Benjamins, Amsterdam/Philadelphia.

Rakover, S.S. 2011/2012a. A plea for methodological dualism and multi-explanation framework in psychology. *Behavior and Philosophy* 39/40:17–43.

Rakover, S.S. 2011/2012b. Methodological dualism and multi-explanation framework in psychology: Replies to criticisms and further developments. *Behavior and Philosophy* 39/40:107–125.

Robb, D. and Heil, J. 2014. Mental causation. In E.N. Zalta (editor), *The Stanford encyclopedia of philosophy*, spring 2014 edition. http://plato.stanford.edu/archives/spr2014/entries/mental-causation/

Rosenberg, A. 1988. *Philosophy of social science*. Westview Press, Boulder, CO.

Salmon, W.C. 1989. *Four decades of scientific explanation*. University of Minneapolis Press, Minneapolis.

Samuels, R., Stich, S., and Faucher, L. 2004. *Reason and rationality*. Handbook of epistemology.

Schueler, G.F. 2003. *Reasons and purposes: Human rationality and the teleological explanation of action*. Oxford University Press.

Searle, J.R. 1980. Minds, brains and programs. *The Behavioral and Brain Sciences* 3:417–457.

Searle, J.R. 1990. Is the brain's mind a computer program? *Scientific American* 262(1):20–25.

Sun, R. and Franklin, S. 2007. *Computational models of consciousness: A taxonomy and some examples*. Cambridge University Press.

Swartz, N. 1985. *The concept of physical law*. Cambridge University Press.

Utall, W. 2014. Are neuroreductionist explanations of cognition possible? *Behavior and Philosophy* 41:37–64.

van Fraasen, B.C. 1980. *The scientific image*. Clarendon Press, Oxford.

von Wright, G. 1971. *Explanation and understanding*. Routledge and Kegan Paul, London.

Weinert, F. 1995. Laws of nature—laws of science. In F. Weinert (editor), *Laws of nature: Essays on the philosophical, scientific and historical dimensions*, de Gruyer, Berlin and New York.

Woodward, J. 2000. Explanation and invariance in the special sciences. *British Journal for the Philosophy of Science* 51:197–254.

Woodward, J. 2002. Explanation. In P. Machamer and M. Silberstein (editors), *The Blackwell guide to the philosophy of science*, Blackwell, Oxford.

Woodward, J. 2003. *Making things happen: A theory of causal explanation.* Oxford University Press.

Yoo, J. 2015. Mental causation. In *The Internet Encyclopedia of Philosophy.* http://www.iep.utm.edu/

13

Biology: Gambler's Epistemology vs. Insistence on Impractical Naturalism—The Unwitting Half-Billion Dollar Wager by the NIH Against Evolutionary Theory

SALVADOR CORDOVA

Millennium Analytics

Abstract

The 2015 Nobel Prize winner in chemistry, Aziz Sancar, may have unwittingly given life to Paley's watch argument when he used the phrase "Rube Goldbergesque designs" to describe the nano-molecular clocks that provide timing to various processes in the human body. Other Rube Goldbergesque designs have been elucidated by National Institutes of Health (NIH) research initiatives such as the ENCODE and RoadmapEpigenomics projects, which represent approximately a half-billion dollar total investment.

The success of NIH initiatives and various other projects has drawn a bizarre reaction from some methodological naturalists such as evolutionary biologist Dan Graur who said in 2012 "If ENCODE is right, evolution is wrong." Graur's comment is reminiscent of Haeckel who said in 1876, "If we do not accept the hypothesis of spontaneous generation, then at this one point in the history of evolution we must have recourse to the miracle of a supernatural creation."

An unconventional approach called "gambler's epistemology" is introduced as a perspective to clarify why naturalism should not be equated with science. Gambler's epistemology, with its reliance on the notion of mathematical expectation, shows that the intuitive perception that "life is a miracle" is not rooted in after-the-fact, ad-hoc probabilities, but is consistent with standard practice in science. Thus without formally settling the question of whether God or supernatural entities actually exist, Haeckel's unwitting assertion that the emergence of life must be of miraculous origin is at least closer to the truth, statistically speaking.

Gambler's epistemology also shows that applying reward-to-risk analyses such as that seen in the professional investment and gambling world might be a better practical guide in committing financial and human resources to scientific exploration than the enforcement of unspoken creeds of impractical naturalism that may actually be detrimental to scientific discovery.

1   Introduction

Though it may be intuitively satisfying to attempt explanations of various phenomena in terms of accessible and repeatable mechanisms such as those deduced via the scientific method, there may be physical phenomena whose explanations escape such reproducibility. One of the most prominent examples is the emergence of life. Despite the evidence from Pasteur's 1861 experiments refuting spontaneous generation, as late as 1876, Haeckel maintained that it was an ordinary and common occurrence for life to emerge spontaneously from non-living matter. This belief was epitomized by his 1876 statement, "If we do not accept the hypothesis of spontaneous generation, then at this one point in the history of evolution we must have recourse to the miracle of a supernatural creation" (Haeckel, 1878, pg. 348).

Despite Haeckel and others' insistence, the belief that the structure of life could be easily explained by repeatable mechanisms was falsified experimentally. Even the most basic life forms are so complex and exceptional relative to non-living matter that, even now, some scientists argue that the emergence of the first life on Earth may not be subject to ordinary and repeatable mechanisms as a matter of principle, making this outside of direct scientific explanation (Trevors and Abel, 2004).

There is a lasting tribute to Pasteur's experiments against spontaneous generation by the word "pasteurized" on bottles of milk. The pasteurization process is a testament to the scientifically verified viewpoint that, for all practical purposes, the emergence of life is so exceptional that it is not expected to happen again. If there is a lesson to be learned from Haeckel's flawed views on the emergence of life, it is that insistence on explanations for all phenomena in terms of repeatable mechanisms should not be conflated or equated with scientific understanding.

If science supports the insight that a phenomenon is so exceptional that it appears miraculous (if only statistically speaking, not theologically speaking), then this insight should not be suppressed merely because it conflicts with the claims of naturalism. For example, the origin of the universe and the origin of life are events that are scientifically inferred as real, not repeatable, and highly exceptional. They can be accepted as such even if that challenges the naturalistic viewpoint. Naturalism is based on the implicit creed that all phenomena can be explained by repeatable and ordinary mechanisms. Therefore, whenever naturalists encounter phenomena that are not reducible to ordinary and repeatable mechanisms, they must maintain their assumptions as faith-based assertions without formal proof.

Supposing for the sake of argument that no God or supernatural forces exist, it would then be hypothetically possible that a phenomenon could be so exceptional and singular that it cannot be reproduced, and hence, it would, as a matter of principle, be outside direct scientific verification. But this raises a philosophical question (that is relevant but unfortunately beyond the scope of this essay), "At what point would a so-called natural phenomenon be so exceptional that it is statistically indistinguishable from a miracle of supernatural origin?"

Extending Pasteur's law of biogenesis that "life comes from life," one might claim on experimental grounds alone that it is quite reasonable to assume animals emerge from animals, plants from plants, eukaryotes from eukaryotes, etc. But these experimental observations would suggest immutability of certain forms (Denton, 1986), and, consequently, macroevolutionary steps are the result of exceptional, rather than typical, events. The claims of macroevolution still remain a matter of inference (and some would say imaginative storytelling) rather than physical experiment (McHugh, 2005). So, in addition to the origin of the universe and the origin of life, one might include the origin of complexity in novel biological forms to be the result of unique, exceptional, and non-repeatable events. If experimental science cannot practically confirm macroevolutionary transitions, evolutionary biology's status as a scientific discipline might be deemed dubious, at least relative to physics and chemistry. As evolutionary biologist Jerry Coyne said, "In science's pecking order, evolutionary biology lurks somewhere near the bottom far closer to phrenology than to physics" (Coyne, 2000).

2   Gambler's Epistemology

An unconventional, but hopefully fruitful, perspective in framing the issue of naturalism vs. the scientific method is the perspective often adopted by professional gamblers and investors. They are dealing in realms where uncertainty is the norm in decision-making. For the purposes of this paper, this informal perspective will be labeled "gambler's epistemology." Gambler's epistemology is neither formally codified nor used as a term explicitly in the gambling and investment world, but coined for the purposes of this essay as a label for a body of principles used by skilled gamblers and investment managers. Rather than offer a strict definition of gambler's epistemology, it will suffice to mention some of the elements of this epistemology that are relevant to naturalism and the scientific method.

The principles of gambler's epistemology are listed in numerous books by specialists even if the term "gambler's epistemology" is not. Therefore, to begin this investigation, it might be helpful to highlight some of the most successful practitioners of this epistemology. Edward O. Thorp was a professor of mathematics at MIT and the author of several books such as *The Mathematics of Gambling (Thorp, 1984)*, *Beat the Dealer (Thorp, 1966)*, and *Beat the Market, A Scientific Stock Market System (Thorp, 1967)*. He teamed up with Claude Shannon (the famous pioneer of information theory) during his successful foray into using computer and mathematical analyses to develop techniques to win money from casinos and Wall Street. Thorp, with Shannon's support, published his first work on gambling in the *Proceedings of the National Academy of Sciences* in 1961 (Poundston, 2005; Thorp, 1961). Thorp made hundreds of millions of dollars after starting an investment fund that applied his theories. His pupil, Bill Gross, went on to manage a trillion dollar hedge fund (Patterson, 2008).

Many decisions in the realm of human affairs are made with far fewer facts available than the decision-makers would like. In the world of successful gambling and hedge fund investment, uncertainty is the order of the day. But uncertainty in one dimension does not necessarily imply uncertainty in another dimension. In fact, maximizing uncertainty in one aspect of a system can lead to near certainty about another aspect of a system. This paradox about reality has been exploited profitably in the business world particularly in casino gambling and casino management, the insurance industry, and investment arbitrage.

The ability to obtain near certainty about one aspect of a system despite uncertainty about another aspect of the same system is easily illustrated by applying the law of large numbers to a system of 500 fair coins. If we take 500 fair coins, place them in a jar, shake them vigorously, and then pour them out on a table, we will induce maximum uncertainty in the heads-tails configuration of each coin. But given the binomial distribution, we can be practically certain the coins will not be 100% heads.

Fundamental to the law of large numbers is the notion of mathematical expectation, or expected value, which was pioneered by mathematician Blaise Pascal in the middle of the seventeenth century. Expected value is the expected average of many outcomes or the average behavior of a system that is composed of many parts. For example, the expected number of coins that are heads in a large system of randomly shaken coins is 50%. The law of large numbers says that deviations from this expectation would be increasingly exceptional the larger the deviation. For 500 fair coins, 100% heads would be an astronomically large deviation from the expectation of 50% heads from a random (i.e., uncertainty maximizing) process. One hundred per cent heads using fair coins and a randomizing process, such as shaking them in a jar and pouring them on a table, would be a statistical "miracle."

Though it would take some work to rigorously formulate the notions of average verses exceptional types of outcomes for a deck of cards stirred by a tornado, suffice it to say a tornado is not mathematically and physically expected to spontaneously assemble a house of cards. If we happened upon a house of cards, we would expect that it wasn't the result of an uncertainty maximizing process like a tornado. The perception that a house of cards is a special configuration relative to a random arrangement of cards is not due to some after-the-fact projection of a pattern by our mind, but can be derived from physical and mathematical principles of expectation. If one were to play the devil's advocate and argue that mathematical expectation is itself an imaginary construct and there are not in actuality any special configurations in the universe (e.g., a house of cards or life), then one would have to abandon all scientific inferences that are based on the notion of expected results. This would effectively dispose much of science.

As scientists learn more about complexity of life, the high specificity of its components, and their connections to each other, it becomes increasingly more difficult to argue that life is the result of ordinary or typical events in a way that makes mechanical sense. It is much like arguing a 747 can be assembled by a tornado passing through a junkyard (Hoyle, 1981). This applies not only to the origin of life problem but also to creating functional biochemical systems that require the emergence of numerous interconnected parts (Behe, 2006).

Finally, the notion of expected value can be applied to wagering and investment decisions such that the best investment is chosen by wagering on a choice that has the highest expected value in terms of payoff. This expected value is calculated by weighing the potential reward on a bet by the probability that the bet will win.

For example, if there is a million dollar payoff for being right, but only a 1% chance of being right, the expected value payoff is $1,000,000 \times 1\% = \$10,000$. Whereas if there is zero payoff for being right and a 99% chance of being right, the expected payoff is $\$0.00 \times 99\% = \$0.00$. If the cost of placing a wager is a mere $100, over many trials, it is better to wager $100 on the %1 long shot that offers the greatest reward than on the 99% certain bet that offers zero reward.

A business executive by the name of Don Johnson was able to win millions from casinos in part by exploiting marketing coupons and rebates that were structured with comparable odds and payoffs as illustrated in the previous paragraph (Bowden, 2012). The trick, of course, was for Johnson to find and negotiate absurdly favorable terms for himself against the casinos. Thorp and Gross used similar strategies to construct their highly successful casino and hedge funds strategies. Pascal himself extrapolated his wagering ideas to the realm of the theological in his controversial claim known as "Pascal's Wager" over the existence of the Christian God (Lataste, 1911).

Because of the law of large numbers, an investment strategy based on selecting investments with the highest expected value payoff will yield on average the best return over a large number of trials. This procedure has been highly effective in business contexts. There is abundant literature on the topic so it will not be covered here in detail except as it applies to the question of investing resources into scientific research on the complexity of life.

3 Evolutionary Biologists vs. the National Institutes of Health: The Half-Billion Dollar Exploration of the Epigenome

Complexity, or the exceptional quality of physical systems, is not an artifact of our imagination, but can be derived from physical and mathematical analyses alone. The origin-of-life problem is a prime example of how the hope of naturalism to explain all phenomena in terms of ordinary and repeatable mechanisms was dashed. But less well-known is the recent discovery of large-scale complexity in the epigenome of life. This is challenging explanations made solely in terms of ordinary and repeatable mechanisms.

A set of projects known as ENCODE and RoadmapEpigenomics, which commands a combined research budget exceeding half a billion dollars, is at the forefront of efforts by the National Institutes of Health (NIH) to explore the genome and epigenome. This research has contributed to the development of FDA-approved treatments such as histone deacetylase inhibitors for the corrupted epigenomes resulting in rare cancers (US Food and Drug Administration, 2015).

Beyond the benefits to medical science, the insights gained from the ENCODE and RoadmapEpigenomics projects have led the projects' researchers to go out on a limb and pronounce that they believe the genome is at least 80% functional. Their declaration was summarized by the 2012 headline in the journal *Science*:

> 'ENCODE Project Writes Eulogy for Junk DNA' - This week, 30 research papers, including six in Nature and additional papers published online by Science, sound the death knell for the idea that our DNA is mostly littered with useless bases.
>
> (Pennisi, 2012)

When these researchers declared that the genome was ten times more functionally complex than previous estimates by certain evolutionary biologists, this induced a reaction epitomized by evolutionary biologist Dan Graur who said, "If ENCODE is right, evolution is wrong" (Luskin, 2014). Dan Graur also offered these thoughts:

> the evolution-free philosophy of ENCODE has not started in 2012...the wannabe ignoramuses, self-promoting bureaucrats, and ol' fashion crooks of ENCODE are protected from criticism and penalties for cheating by the person who gives them the money. Thus, they can continue to take as much money from the public as their pockets would hold, and in return they will continue to produce large piles of excrement that are hungrily consumed by gullible journalists who double as Science editor.
>
> (Graur, 2015)

Graur is a professor at the University of Houston and fellow of the American Association for the Advancement of Science (AAAS). Graur and several co-authors, with the full sanction and cooperation of several fellow evolutionary biologists, published the paper, "On the Immortality of Television Sets: 'Function' in the Human Genome According to the Evolution-Free Gospel of ENCODE" (Graur, Zheng, Price, Azevedo, Zufall, and Elhai, 2013). The paper passed peer-review by other evolutionary biologists despite its over-the-top tone that bordered on name-calling. Notably, several in the scientific community objected to the paper's overtly hostile and unprofessional tone saying it was ill-suited for science and scholarship (Rehman, 2013). Graur's shrill rhetoric also inspired a reporter for *Science* to refer to him as "The Vigilante" (Bhattacharjee, 2014).

Despite Graur's tone, he has supporters among evolutionary biologists and population geneticists. One of the world's most respected theoretical geneticists, Joseph (Joe) Felsenstein, authored *Theoretical Evolutionary Genetics*, the gold standard graduate textbook in evolutionary genetics. In the book, Felsenstein explicitly mentions the ENCODE project and why its claims are at odds with the mathematics of evolutionary genetics (Felsenstein, 2016, pg. 155), which would imply Graur's assertion that "If ENCODE is right, evolution is wrong".[1]

Thus, in the present day, we are in a situation where textbooks on evolutionary genetics are openly in conflict with the claims of highly respected laboratory researchers commissioned by the NIH. There would appear to be some uncertainty in deciding where to focus research efforts in the face of unresolved questions over evolution and the results of ENCODE. But such situations are tailor-made for applying gambler's epistemology.

4 ENCODE, RoadmapEpigenomics, E4

Subsequent to the success of the multibillion dollar Human Genome Project, which enumerated the DNA sequences in the human genome and which was completed in 2003, there was still a question as to how the individual parts of the genome worked. The head of the Human Genome Project and now the current director of the NIH, Francis Collins, predicted it would take centuries to understand how each part of the genome works (Collins, 2007). Among the first steps into this endeavor was the NIH ENCODE project, whose mission was to begin cataloging the individual parts of the genome and identify their roles.

The ENCODE project, beginning in 2003, commanded a budget of 288 million dollars (Bhattacharjee, 2014). The RoadmapEpigenomics project, which began in 2008, has a budget of 300 million dollars (Pennisi, 2015). There is a peripherally

---

[1]The appendix lays out a simplified description of Felsenstein and Graur's arguments, which are (ironically and for totally different reasons) supported by creationists like respected Cornell geneticist John Sanford.

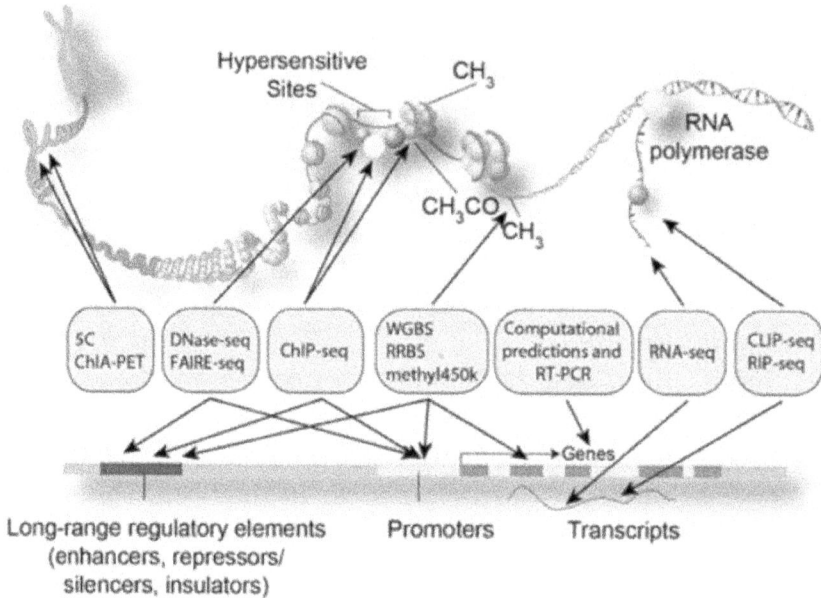

Figure 13.1: A small sampling of the experiments conducted by ENCODE on a stretch of DNA (the multi-colored bar toward the bottom). The gray bubbles represent classes of experiments. Many of the experiments, such as WGBS (whole genome bisulfate sequencing), RRBS, methyl450k, ChiP-seq, and RNA-seq, are relevant to exploring the human epigenome.

Image credits: Darryl Leja (NHGRI), Ian Dunham (EBI), Michael Pazin (NHGRI), public domain photo funded by US Government.

related project that is in the planning stages called E4 (Enabling Exploration of the Eukaryotic Epitranscriptome) that has a projected budget of 205 million (Satterlee, 2015).

The ENCODE project developed many experimental techniques and established databases that are now being used in the follow-up RoadmapEpigenomics project. There are about forty classes of experiments performed by ENCODE, some of which are depicted in Figure 13.1 (Davidson, T.Chan, Sloan, L.Hong, S.Malladi, Rowe, Strattan, Ho, Podduturi, Hitz, Tanaka, Lee, Simison, Kent, and Cherry, 2015).

The experimental findings of the ENCODE project startled the researchers since it suggested substantially more of the genome was functional (80% or more) than predicted by evolutionary theorists (less than 10%) (Rands, Meader, Ponting, and Lunter, 2015). This functionality includes DNA's involvement in a conceptual entity

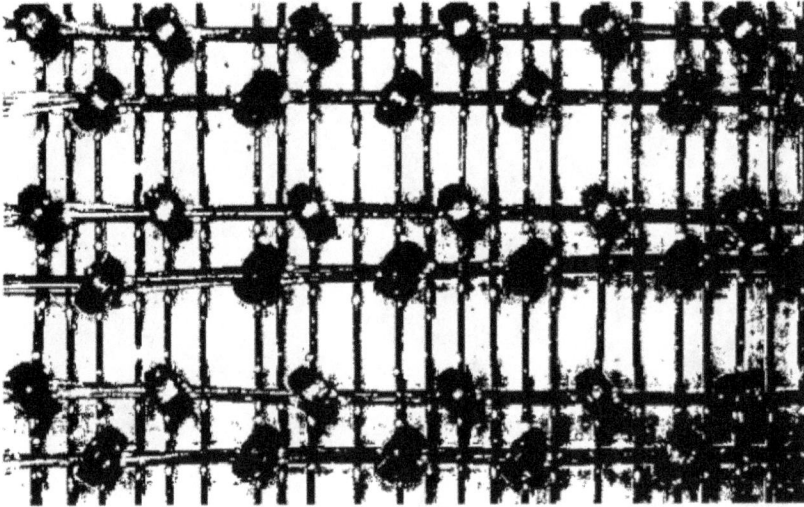

Figure 13.2: Primitive man-made random access memory (RAM). Notice the "beads on a string" structure. Each "bead" is where a single bit of memory is stored. Photo has been decolorized with added graininess to allow a more direct comparison with pictures of the nucleosomes generated from an electron microscope.

Image credit: Steve Jurvetson from Menlo Park, USA, through Wikimedia commons, available under the Creative Commons Attribution 2.0 Generic license. https://commons.wikimedia.org/wiki/File:Primitive_ Magnetic_Memory.jpg

known as the epigenome.

DNA is widely viewed as read only memory (ROM), but the ENCODE project advanced the emerging view that DNA also acts as a component of cellular random access memory (RAM). Figure 13.2 and Figures 13.3 and 13.4 show an amusing coincidence of a "beads on a string" structure that appears in man-made RAM as well as biological RAM.[2]

The term "epigenome" is unfortunately, and disparagingly, used in association with Larmarkian and Lysenkoist ideas, but it means something almost totally different to those who view the epigenome as a complex entity that is necessary for the

---

[2]The structure here is referred to as "beads on a string" where the "beads" are the histone/nucleosomes and the "string" is the DNA connecting the beads. The *Stem Cell Handbook* refers to this complex as part of the "random access memory" of the cell (Papatsenko, Xu, Ma'ayan, and Lemischka, 2013, pg. 71). Amazingly, even though the chemical and physical mechanisms of memory in this biological RAM are different than the man-made RAM depicted in Figure 13.2, they coincidentally conform to a "beads on a string" architecture.

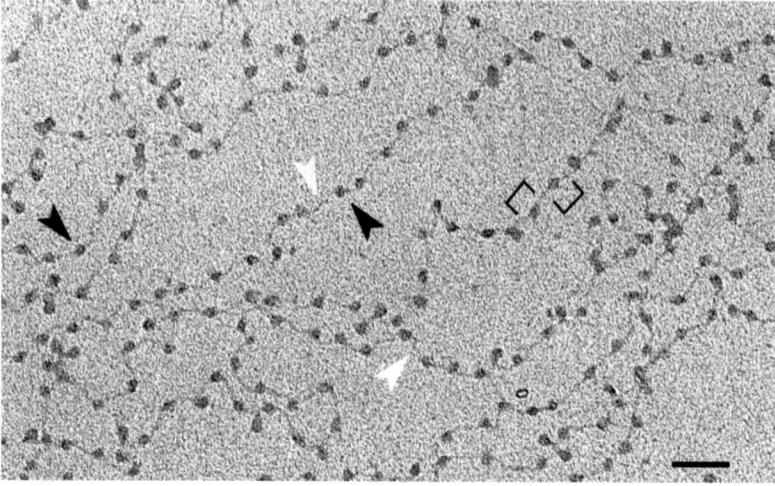

Figure 13.3:   Biological RAM. Electron micrograph of histone/nucleosome complexes of DNA.

Image credit: Chris Woodcock, available under the Creative Commons Attribution-Share Alike 3.0 Unported license.
https://commons.wikimedia.org/wiki/File:Chromatin_
nucleofilaments_(detail).png

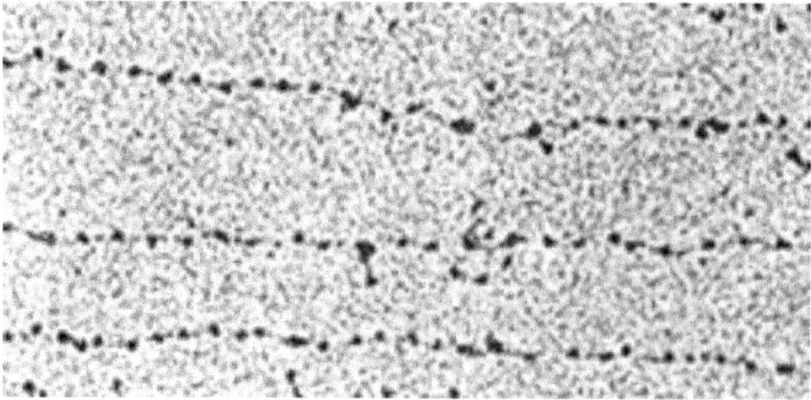

Figure 13.4:   Biological RAM. Electron micrograph of histone/nucleosome complexes of DNA

Image credit: Dr. B. Hamkalo, University of California, Irvine as published by Cold Spring Harbor Laboratory DNA Learning Center, available under a Creative Commons Attribution-Noncommercial-No Derivative Works 3.0 United States License.
https://www.dnalc.org/view/16636-Gallery-29-Electron-micrograph-of-chromatin.html

development of somatic cells in a multicellular organism. Roughly speaking, there may be one genome, but there are as many epigenomes in an adult human as there are cells.

Though the definition epigenome is in constant flux, a segment of researchers generally define the epigenome as including methyl modifications to the DNA itself, chemical modifications to the histones which the DNA wraps around, and even the non-coding RNAs that are involved in cellular operations (Chial and Akst, 2012). The word "epigenetic" refers to isolated parts of the epigenome, and the epigenome refers to the sum total of epigenetic parts. The ENCODE and RoadmapEpigenomics projects are generally sympathetic to the definition of the epigenome that emphasizes methyl modifications to DNA and chemical modifications to histones.

A more accurate term instead of "epigenome" might be "chromatin modifications." For the purposes of this paper, unless otherwise stated, "epigenome" will mean "chromatin modifications."[3] Each cell has a different set of chromatin modifications. Thus there is the potential for an adult with 100 trillion cells to have 100 trillion epigenomes.

The *Stem Cells Handbook* considers the genome analogous to ROM and the epigenome analogous to RAM (Papatsenko et al., 2013). As an academic exercise, one can attempt to count the hypothetical number of possible chemical states in the combined collection of 100 trillion epigenomes in an adult human. One way to express the number of states, since it is astronomically large, is in terms of Shannon information bits. Given that there are about 100 trillion cells in the adult human, 16.5 million nucleosomes per cell, and at least 40 bits of information per nucleosome (Cota, Shafa, and Rancourt, 2013, see also Figures 13.5 and 13.6), a back-of-the-envelope calculation yields an approximate total RAM in an adult human to be on the order of sextillion (10^{21}) bits. Some of this RAM is believed to be utilized in the brain for learning and cognition, the body for self-healing and development, and many yet-to-be discovered functions required to operate the various organs and systems in the body. The ENCODE and RoadmapEpigenomics projects have contributed to our understanding of how this enormous amount of epigenetic RAM is utilized.

In addition to the epigenome, in the last few years, there has emerged the notion of the epitranscriptome, which represents chemical modifications to RNA transcripts (Saletore, Meyer, Korlach, Vilfan, Jaffrey, and Mason, 2012). The degree of complexity needed for a eukaryotic organism to manage this vast amount of information suggests that it is incompatible with current evolutionary genetics.[4] But suffice it to say, if evolutionary genetics cannot explain the complexity of the epigenome and the epitranscriptome, it is not currently (and may never be) feasible to explain biological complexity in terms of repeatable and ordinary mechanisms. Thus it weakens

---

[3]Whatever the labels, epigenomic research commands a large amount of financial interest in the medical community with an estimated therapeutic market of 8 billion dollars in 2017 and is anticipated to receive more funds in the future (Aster, 2010).

[4]The appendix will go into some of the technical details of this inference.

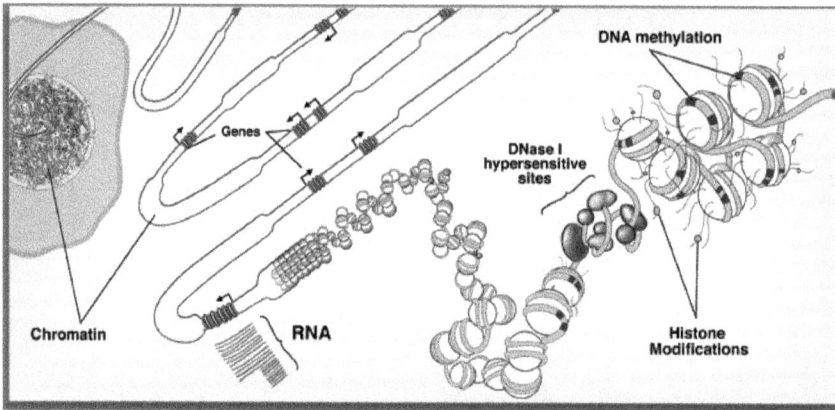

Figure 13.5: A depiction of DNA conceptually uncoiled from the cell nucleus (left) to reveal the "beads on a string" architecture of chromatin. Chromatin is composed of DNA wrapped around histones. The "bead" is called a nucleosome and consists of DNA wrapped around 8 histones. The nucleosomes occur at a frequency of about one every 200 base pairs of DNA.

Image credit: The RoadmapEpigenomics project website of the National Institutes of Health (US Government), public domain.
http://www.roadmapepigenomics.org/

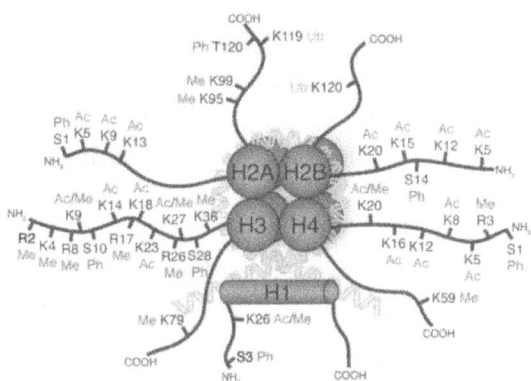

Figure 13.6: A tabulation of the known chemical modifications to histone tails in the DNA nucleosome complexes. Each nucleosome occurs regularly for about every 200 base pairs of DNA. This figure shows 42 possible chemical modifications to the histone core of a nucleosome, but there are likely more modifications to be discovered. "Me" means methylation, "Ac" acetylation, "Ph" phosphorylation, "Ub" ubiquitination. Each modification can be approximated as representing 1 bit of information. In truth, a chemical modification shown in the above diagram can sometimes represent more than one bit of information because in cases such a methylation, there are up to 3 different degrees of methylation (Cota et al., 2013).

Image credit: Cota et al. (2013), available under the Creative Commons Attribution 3.0 Unported license. http://www.intechopen.com/books/pluripotent-stem-cells/stem-cells-and-epigenetic-reprogramming the claims of naturalism to the extent that naturalism denies the existence of highly exceptional processes that would qualify as statistical miracles.

5   Conclusion

Although it is a philosophical question as to what point a phenomenon passes the threshold of being either natural or supernatural, a sufficiently extraordinary set of events might be perceived as indistinguishable from a supernatural miracle even if, hypothetically, there were no God or gods to speak of. The exceptional property of life is illustrated by Haeckel's claim that if the theory of spontaneous generation is false, then the emergence of life would have to be of a miraculous supernatural origin. Questions about God and the existence of the supernatural are outside the scope of this paper, but the resolution of the question of God and the supernatural are not needed to realize that naturalism is not equivalent to science.

The case for naturalism is weakened if a phenomenon exists that hints at the involvement of astronomically exceptional circumstances for it to emerge. It would appear that life is one such phenomenon. If the specialness of life doesn't challenge naturalism, at the very least, it challenges the ability to explain it in terms of ordinary and repeatable processes.

It is understandable that some methodological naturalists find the idea of miraculous-looking complexity in life as incompatible with a naturalistic narrative that insists miraculous events had no role in the emergence of life and its complex features. But such sentiments are speculations, and, though superficially sounding like scientific explanations, such assertions should not be conflated or equated with actual science, and hence investment decisions in committing resources to scientific exploration should not be constrained merely because such explorations have the potential to discover facts that are unfavorable to naturalistic philosophy.

If the conclusions from ENCODE are right and the genome is more functional than evolutionary biologists have argued, but no money is invested in research that is friendly to ENCODE's claims, medical science and the chance to alleviate human suffering risks being permanently compromised. On the other hand, if money is invested to prove that Dan Graur and the evolutionary biologists he represents are right, there will be no benefit to the human medical condition even if they are right. Thus, according to Pascal's wagering theories, in light of these payoffs, and provided there is some small probability that ENCODE is right, money should be wagered on ENCODE, and indeed that is where the money is being wagered by the NIH on behalf of US taxpayers.

Insisting on the truth of naturalism disguised as evolutionary theory could impede scientific progress in the medical sciences if the whims of some evolutionary biologists like Dan Graur are realized. The National Science Foundation (NSF) has invested 170 million dollars in unresolvable evolutionary phylogenies that have been of little or no utility to medical science (Tan and Tomkins, 2015). To date, no therapies based on the 170 million dollar phylogeny project have come to market. In contrast, with help from research projects like ENCODE, epigenetic therapies are already being delivered to patients and more such therapies in the pipeline. Therefore, a gambler's epistemology that seeks to maximize the reward in the face of uncertainty would seem a superior approach versus blind insistence on impractical projects based on naturalism.

Appendix 1: Simplified Explanation of Genetic Entropy and Reasons for Dan Graur's Complaint Against ENCODE

A population can tolerate a certain number of mutations per individual per generation. The tolerable load of mutations is also known as "mutational load." Graur's complaint against ENCODE can be summarized as the problem of mutational load.

Calculations of mutational load for humans was prominently put forward by Hermann Muller who estimated the human genome can tolerate at most one bad mutation per individual per generation (Muller, 1950). Muller won the Nobel Prize for his research into genetic deterioration due to radiation.

If ENCODE is right, the functional genome would be on the order of 3 giga base pairs, and given accepted mutation rates and the size of the functional genome, this would imply that there are on the order of 50 to 100 function-compromising mutations per generation per individual (Graur, 2016). Graur himself explicitly said:

> If 80% of the genome is functional, as trumpeted by ENCODE Project Consortium (2012), then 45-82 deleterious mutations arise per generation. For the human population to maintain its current population size under these conditions, each of us should have on average 3×10^{19} to 5×10^{35} (30,000,000,000,000,000,000 to 500,000,000,000,000,000,000,000,000,000,000,000) children. **This is clearly bonkers**. If the human genome consists mostly of junk and indifferent DNA, i.e., if the vast majority of point mutations are neutral, this absurd situation would not arise. [emphasis added]

Darwin and Spencer asserted "survival of the fittest" as an axiom of nature. But survival of the fittest occurs between siblings and cousins (figuratively speaking) of a generation, not between ancestors and descendants across generations. If, on average, the children are substantially more damaged than their parents, no amount of selecting the best kids among their peers will result in genetic advancement over time. Rather it would be genetic deterioration even though the axiom of "survival of the fittest" held true. A simplified conception of this problem is illustrated in Figure 13.7.

The problems posed by mutational load and other aspects leading to genetic deterioration has been summarized in a book by genetic engineer John Sanford at Cornell (Sanford, 2014). Though Sanford is a creationist, he would likely agree with Graur and the evolutionary biologists, "If ENCODE is right, evolution is wrong."

Appendix 2: Life as a Rube Goldberg Machine

A Rube Goldberg machine is customarily defined as a contraption, invention, device or apparatus that performs a simple task in an indirect, convoluted, and complicated fashion. It is named after American cartoonist and inventor Reuben Garrett Lucius "Rube" Goldberg (1883–1970).

In 1996 Michael Behe in his book *Darwin's Black Box* used notion of Rube Goldberg Machines to describe complex biochemical systems (e.g., blood clotting and vision). However, the term "Rube Goldberg Machine" was overshadowed by his idea of "Irreducible Complexity" (Behe, 2006).

Behe's ideas have had influence on other biologists even though they disagree with his claims of Intelligent Design and his criticisms of Darwinism. A possible hint of Behe's influence on scientific culture is suggested by a description and diagram in the 2010 *Cell and Molecular Biology* textbook by Gerald Karp, which showed a picture of a Rube Goldberg machine with the following caption (Karp, 2009):

> Cellular activities are often analogous to this Rube Goldberg machine in which one event automatically triggers the next event in a reaction sequence.

Additionally, the 2015 Nobel Prize winner in Chemistry, Aziz Sancar, wrote the 2008 paper "The Intelligent Clock and the Rube Goldberg Clock" (Sancar, 2008). In that paper, Sancar uses the phrase "Rube Goldbergesque Designs" to describe the eukaryotic biological clock. Intelligent Design advocates might be tempted to argue Paley's watch is a molecular Rube Goldberg Machine (Evolution News and Views, 2012).

The phrase "Rube Goldberg" has been used both as a term of derision, as in Sancar's case, and as a term of praise, as in Behe's case, for the way biological systems are constructed. Rube Goldberg machines transcend the characterization of "good design" and "bad design" since Rube Goldberg machines are "bad designs" in the sense that they are excessively complex, which increases the fragility of the system, but they are "good designs" in the sense that they showcase the creativity and ability of the designer to balance a design on the edge of functionality and disaster (like a house of cards).

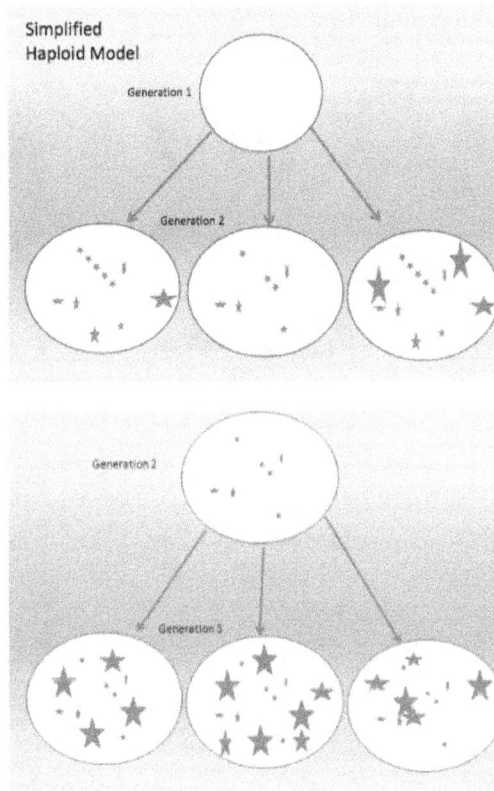

Figure 13.7: Conceptual diagram of inevitable genetic deterioration. The Bubbles represent individuals in a generation and the red stars are detrimental mutations. The parent from generation 1 hypothetically has no mutations. Each of the offspring in generation 2 has several mutations that the parent didn't have. Even supposing the offspring in generation 2 with the least number of mutations is selected to spawn the offspring in generation 3, it will pass on its defects to its children in addition to adding new bad mutations to offspring in generation 3. The number of detrimental mutations increases with each generation. Granted this is a simplified single-parent (haploid) model, but it is provided for conceptual purposes. The more complex models (such as those developed by Muller and others, which leverage the Poisson distribution) arrive at the same essential conclusion. Furthermore their calculations yield a estimate that one bad mutation per individual per generation on average cannot be tolerated by the human species.

Image credit: Salvador Cordova, 2016

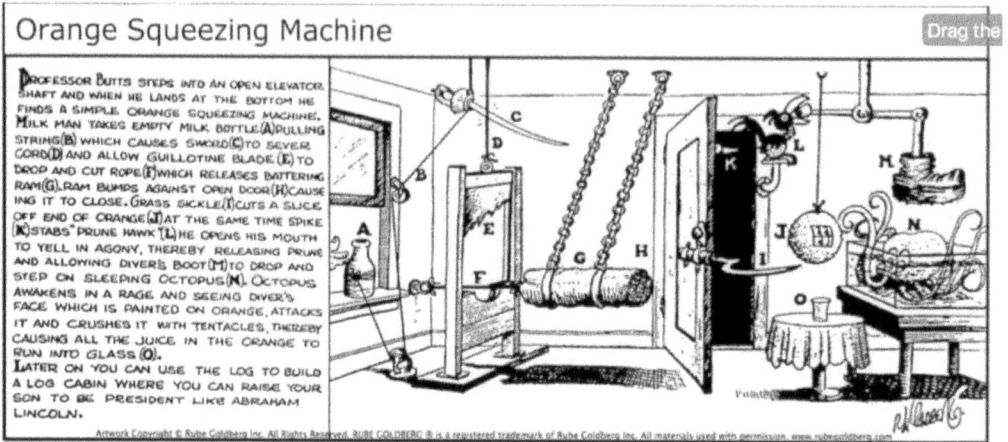

Figure 13.8: An orange squeezing Rube Goldberg machine which was featured in Gerald Karp's 2010 book *Cell and Molecular Biology Textbook* to illustrate cellular activities.

Image credit: Rube Goldberg Inc.

Evolutionary biologists assume that natural selection is sufficient to account for the level of complexity we see in life (Dawkins, 2015). This has never been theoretically or empirically established. Evolutionary biologists have not offered a formal computation of the mathematics and probabilities behind selection of more complex traits from simpler systems. At best the question is unanswered; at worst the assumption that selection frequently selects for complexity is completely wrong.

The peacock's tail and the problem of Rube Goldberg-like extravagance, according to Darwin, made him "sick" (Burkhardt, Browne, Porter, and Richmond, 1993, pg. 140). He reasoned that natural selection should select against the extravagance of such complexity rather than for it. Such extravagant complexity results in the species being more vulnerable and fragile, and thus less likely to survive. He reasoned that there was a mechanism other than natural selection that led to the peacock's tail. Since he rejected special creation, he suggested a theory of sexual selection (whereby mates select for complexity). However, that leaves open the question of what created the extravagance of sexual reproduction in the first place, not to mention that if sexual selection created extravagance that reduced survivability for the species as a whole, then such extravagance would still be selected against by natural selection, even if it was selected for by sexual selection.

According to a PBS documentary on evolution, prior to Darwin's theory, the prevailing view was that biological organisms were created to attest to the Creator's ingenuity to the men who studied these organisms, not for the creature's survival (Tale of the Peacock, 2001). The peacock's tail remains problematic for explaining biology purely in terms of survival, since such excessively complex systems would

Figure 13.9: A peacock's tail

Image credit: Benson Kua from Toronto, Canada. Wikipedia commons. Creative commons license. https://commons.wikimedia.org/wiki/File:Peacock_in_ Toronto.jpg

increase fragility and thus decrease survivability in a competitive environment.

Darwin's theory equivocates on the notion of "selectively favored." Selectively favored in the present does not imply selectively favored in past. Nevertheless, this equivocation seems to be a staple in "proving" selection theory as a credible theory whereby selection evolves toward non-existent traits (Mivart, 1871).

For example, in mammals a functioning insulin-regulated metabolism is a requisite for life but a dysfunctional one results in death. Population geneticists model selection mathematically by attaching an S-coefficient to a trait. Superficially it would seem selection would favor evolution of an insulin-regulated metabolism and hence a favorable S-coefficient should be attached to that trait in population genetic models. But if an ancestral species does not have an existing insulin-regulated metabolism, and if it is critical for life, then the species will go extinct in one generation. Affixing S-coefficients in mathematical models of the past based on traits that are critical to life in the present is thus completely illegitimate. Nevertheless, models that utilize such illegitimate reasoning are put forward as proof of Darwinian theory with phrases in peer-reviewed literature such as "this trait evolved in order to..." These statements are made while failing to recognize that selection cannot select toward non-existent traits and that Darwinian evolution cannot have the foresight to evolve toward some goal.

On the other hand, if the insulin-regulated metabolism is not critical to life, on what basis can it be argued it was selected for in the past? Creatures with insulin-regulated metabolisms in the present might be nothing like creatures without insulin-regulated metabolisms in the past. For such an insulin-regulated metabolism to evolve, it would require numerous parts to appear simultaneously such as the insulin molecule itself, a means of manufacturing insulin (emergence of new beta cell types), a means of regulating insulin (feedback mechanisms), and a means of responding to insulin (appropriate tyrosine kinase receptors).

Since the likelihood of simultaneous appearances of requisite parts seems astronomically remote, evolutionary biologists postulate co-option (exaptation) whereby these parts were first used by the organism for some other purpose. But this is purely speculative. Even supposing the parts were available, there is the further problem of actually evolving instructions to assemble and utilize the parts.

Supposing someone is given all the necessary characters to solve a forty-character password where each character is unique, it would still be a challenge to solve the password even though all the available characters are known (analogous to a co-option scenario). The chances of solving such a password in one try would be 1-out-of-40 factorial (or 8×10^{47}). An evolutionary algorithm cannot solve a complex password without knowing the actual password in advance. There is no feedback indicating that one is getting closer to a solution with each trial. In like manner, for complex all-or-nothing Rube Goldberg-type systems, selection can't select toward the individual parts since there would be no feedback indicating that one variation is closer to success than another.

In fact, sometimes a half-formed organ or system is worse for the organism than no organ at all. Hence, selection would, in general, select against the formation of novel Rube Goldberg-like complexities. As paleontologist Stephen Jay Gould said, "What good is half a jaw or half a wing?" (Gould, 1980).

Darwin's claims about natural selection could be considered rhetorical false advertising to the extent that it is contrary to experimental and observational evidence, which would make his proposed mechanism un-natural selection. Darwin's views are certainly not what happen naturally in the present day as evidenced by the fact that increased selection pressure on ecological systems leads, on average, to the extinction of complex multicellular forms rather than the emergence of them. For example, it is widely acknowledged that under the increased selection pressure from human ecological intrusion, birds and other complex species are going extinct faster than they are being replaced by new complex forms (Kluger, 2014).

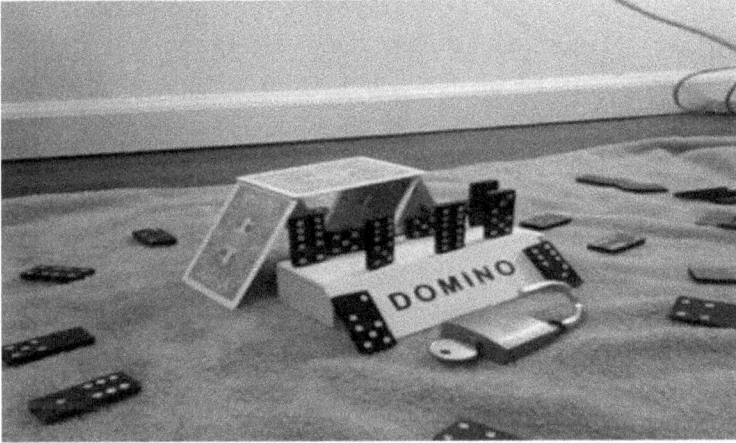

Figure 13.10: A small house of cards behind dominoes standing on a small wooden box behind a lock and key. The house of cards and dominoes illustrate systems of objects that cannot be produced by an uncertainty maximizing process such as a tornado or similar process that affixes random orientations and positions to the objects. The lock and key are included in the photo since it is important to understanding that even though there are an infinite number of ways to make a lock and key, it doesn't imply a high probability that a working lock and key will emerge from random processes. Rather than being high, the probability is remote.

Image credit: Salvador Cordova. Photo of dominoes, cards, lock and key on towel. June 14, 2016.

Appendix 3: Rube Goldbergesque Designs, Specificity, and Complexity

A typical objection to the probability arguments put forward by Intelligent Design proponents is that the probability calculations they advocate are after-the-fact calculations, and are therefore illegitimate. For example, any random shuffle of a deck of cards will yield an astronomically rare sequence that occurs 1-out-of-52 factorial (approximately 8×10^{67}) times. So each random possible sequence of cards is astronomically remote, hence opponents of Intelligent Design would argue any probability calculation about a sequence of cards cannot be used to argue one sequence is more special than any other. They extend the same sort of objections to the emergence and complexity of life.

However the specialness of one sequence is not due to the improbability of a sequence, but how far from mathematical expectation a sequence is. Earlier, it was mentioned that fair coins that turned up 100% heads is maximally far from the expectation of 50% heads. Additionally, suppose we found all the cards belonging to the red suits at the top of the deck and all cards belonging to the black suits at the bottom. The expected value for the first 26 cards at the top of the deck is approximately 50% red cards and 50% black cards, whereas 100% red cards at the top of the deck is farthest from expectation with odds of 1-out-of-5 $\times 10^{14}$.[5] Based on textbook math, the "all red-suited cards at the top" configuration is an exceptional configuration. The symbolic properties (red and black suits) are decoupled from the physical properties, which leads to the possibility that certain special symbolic sequences are not practically explicable by random physical processes, but rather by processes that defy natural randomizing tendencies.

To argue that "we just made all-red special in our minds, but it really isn't" would seem like a last resort attempt to argue the "all red suited cards at the top" of the deck configuration is not special. But to argue there is ultimately nothing exceptional in the universe, that "special" or "exceptional" is rooted in our imagination, would be to undo the foundations of probability theory and much of the science that depends on it.

Independent of the question of whether something is intelligently designed, the question of the existence of exceptional configurations can still be asserted. Living organisms are exceptional chemical configurations based on theory and experiment (e.g., Pasteur's experiments). Our perception of the specialness of life is not a consequence of after-the-fact probabilities or seeing faces in clouds.

Various system parts can be said to have high specificity if they cannot tolerate much change or perturbation without the salient nature of the system breaking down. In Figure 13.9, systems with high specificity are illustrated by the highly specific orientation and position of cards required to put them into a house configuration

---

[5]The calculation is $26/52 \times 25/51... 1/27 = 26!/(52!/26!) = 5 \times 10^{14}$.

(versus lying flat), by dominoes standing on their edges, and by the matching a key to a lock. Random orientations and positions (non- specific configurations) of cards will not result in systems, such as a house of cards, even though there might be an infinite number of ways to make a house, or a village, of cards. Similar considerations apply for the dominoes, lock-and-key systems, and the origin of life.

A system composed of many parts can be said to be complex. Some might call a system possessing both large amounts of specificity and complexity as possessing "specified complexity," but because the phrase "specified complexity" has so many conflicting definitions (some involving information theory, see Specified Complexity, 2016) the term is de-emphasized in this paper.

Instead, the notion of "Rube Goldbergesque Design" is suggested as more descriptive of the nature of biological complexity that is in a highly specified state. It is this class of extravagant complexity that bothered Darwin in the past and bothers evolutionary biologists in the present because such designs would be selected against rather than selected for owing to the fact that greater specificity and complexity (like a larger house of cards) is more vulnerable to failure.

Darwin argued that his theory of natural selection explains the emergence of complexity in biology, but his theory is not what is observed in nature. Even by his own admission as symbolized by the peacock's tail, extravagant Rube Goldbergesque complexity present in life would actually be selected against rather than for.

Framing the spontaneous generation debate in terms of Natural vs. Supernatural or in terms of Intelligent Designs vs. Mindless Design muddles the more basic scientific question. The basic scientific question is whether life is a typical or exceptional chemical configuration. Life is an exceptional chemical configuration, astronomically so. Impractical naturalism is not comfortable with phenomena that hint of events so singular that they would be indistinguishable from miracles. Thus when faced with the fact of an exceptional phenomenon, such as the mechanical complexities of life, proponents of naturalism often try to argue something isn't that complex after all.

It is no surprise, therefore, that evolutionary biologists who are also naturalists are often inclined to insist biological systems are not that complex and that complexity is an illusion. They argue that the convoluted and apparently clumsy ways living things go about their business is evidence against Intelligent Design and in favor of naturalistic evolution. But convoluted mechanisms could just as well be interpreted as Rube Goldbergesque designs, and Rube Goldbergesque designs in nature, like the peacock's tail, could just as easily argue for Intelligent Design as against it. But in any case, it is very hard to argue any Rube Goldbergesque design, be it God-made or nature-made, would be a phenomenon consistent with mathematical and physical expectations of something occurring naturally.

ENCODE and research into epigenomics has uncovered several biological Rube Goldberg machines that epitomize specificity and complexity. One example of such a Rube Goldberg machine is described in the Appendix 4.

Appendix 4: Example of a Biological Rube Goldberg Machine Involving non-Coding RNA

Non-Coding RNAs were widely viewed as mostly junk but those views are changing. There is still significant debate about the characterization and extent of junk in DNA and RNA. One example of a non-coding RNA that was thought to be junk and then was discovered to be functional was the HOTAIR lncRNA. The HOTAIR lncRNA was given that name because the researchers joked that if the lncRNA molecule they were studying turned out to be junk, then their hypothesis would be a bunch of "hotair" (Zimmer, 2015).

John Rinn discovered that HOTAIR lncRNA originated from Chromosome 12 and sails to Chromosome 2 by the winds of Brownian motion. HOTAIR writes modifications to Chromosome 2 on the Histone Random Access Memory (RAM) by recruiting the PRC2 polycomb repression complex (Margueron, 2014). This marking on the RAM of each skin cell causes skin at the soles of the feet to be different from skin on the eyelids. Since Rinn's discovery, HOTAIR has been found to interact with DNA in other chromosomes. HOTAIR's discovery inspired research into the roles of other such non-coding RNAs.

The system involving HOTAIR can be said to be a Rube Goldbergesque design in that a very complex ritual of tightly specified parts is involved in carrying out a task of gene regulation whereby DNA from one chromosome regulates DNA on another chromosome which regulates the differential development of skin cells.

Acknowledgements

Many thanks to John Sanford and the FMS foundation and Jonathan Bartlett and the Blyth Institute for making this paper possible. Thanks also to the Evolutionary Informatics Lab for sponsoring the conference where the elements of this paper were presented.

References

Aster, N. 2010. *Epigenetics Therapeutics to 2017: High Market Potential for Epigenetics Drugs*. Market Publishers.

Behe, M. 2006. *Darwin's Black Box*. Free Press.

Bhattacharjee, Y. 2014. The vigilante. *Science* 343:1306–1309.

Bowden, M. 2012. *The Man Who Broke Atlantic City*. The Atlantic.
http://www.theatlantic.com/magazine/archive/2012/04/the-man-who-broke-atlantic-city/308900/

Burkhardt, F., Browne, J., Porter, D.M., and Richmond, M. 1993. *The Correspondence of Charles Darwin*, volume 8. Cambridge University Press, Cambridge.

Chial, H. and Akst, J. 2012. Epigenetics. *Scitable by Nature Education* .
http://www.nature.com/scitable/spotlight/epigenetics-26097411

Collins, F. 2007. *The Language of God*. Free Press.

Cota, P., Shafa, M., and Rancourt, D.E. 2013. Stem cells and epigenetic reprogramming. In D. Bhartiya and N. Lenko (editors), *Pluripotent Stem Cells*, Intech.

Coyne, J. 2000. Of vice and men: The fairy tales of evolutionary psychology. *The New Republic* 222(14):27–34.

Davidson, J.M., T.Chan, E., Sloan, C.A., L.Hong, E., S.Malladi, V., Rowe, L.D., Strattan, J.S., Ho, M., Podduturi, N.R., Hitz, B.C., Tanaka, F., Lee, B.J., Simison, M., Kent, W.J., and Cherry, J.M. 2015. The role of the ENCODE data coordination center. In *ENCODE 2015: Research Applications and Users Meeting*.

Dawkins, R. 2015. *The Blind Watchmaker*. W. W. Norton & Company.

Denton, M. 1986. *Evolution: A Theory in Crisis*. Adler & Adler.

Evolution News and Views. 2012. Do biological clocks revive William Paley's design argument? .
http://www.evolutionnews.org/2012/06/do_biological_c060411.html

Felsenstein, J. 2016. *Theoretical Evolutionary Genetics*.
http://evolution.gs.washington.edu/pgbook/pgbook.pdf

Gould, S.J. 1980. The return of hopeful monsters. *Natural History* 86:22–30.

Graur, D. 2015. Evolution-free genomics: A 1964 paper as a prelude to #encode_nih. *Judge Starling (Dan Graur's Website)* .
http://judgestarling.tumblr.com/post/120959594801/evolution-free-genomics-a-1964-paper-as-a-prelude

Graur, D. 2016. Rubbish DNA. *Arxiv* .
https://arxiv.org/ftp/arxiv/papers/1601/1601.06047.pdf

Graur, D., Zheng, Y., Price, N., Azevedo, R.B., Zufall, R.A., and Elhai, E. 2013. On the immortality of television sets: 'function' in the human genome according to the evolution-free gospel of ENCODE. *Genome Biology and Evolution* 5(3):578–590.

Haeckel, E. 1878. *The History of Creation.*

Hoyle, F. 1981. Hoyle on evolution. *Nature* 294(5837):105.

Karp, G. 2009. *Cell and Molecular Biology.* Wiley.

Kluger, J. 2014. The sixth great extinction. *Time Magazine* .

Lataste, J. 1911. Blaise Pascal. In *The Catholic Encyclopedia*, Robert Appleton Company, New York.
http://www.catholic.com/encyclopedia/blaise-pascal

Luskin, C. 2014. The ENCODE Embroilment, Part I. *Salvo Magazine* .

Margueron, R. 2014. The polycomb complex PRC2 and its mark in life. *Nature* 469:343–349.

McHugh, P.R. 2005. Teaching darwin: Why we're still fighting about biology textbook. *The Weekly Standard* .
http://www.weeklystandard.com/teaching-darwin/article/6586

Mivart, G.J. 1871. The incompetency of 'natural selection' to account for the incipient stages of useful function. In *On the Genesis of Species*, Macmillan and Company.
http://www.gutenberg.org/ebooks/20818

Muller, H.J. 1950. Our load of mutations. *The American Journal of Human Genetics* 2(2).

Papatsenko, D., Xu, H., Ma'ayan, A., and Lemischka, I. 2013. Quantitative approaches to model pluripotency and differentiation in stem cells. In S. Sell (editor), *Stem Cells Handbook*, pp. 59–74, Springer New York.

Patterson, S. 2008. Old pros size up the game: Thorp and Pimco's Gross open up on dangers of over-betting, how to play the bond market. *Wall Street Journal* .

Pennisi, E. 2012. ENCODE project writes eulogy for junk DNA. *Science* 337(6099):1159–1161.

Pennisi, E. 2015. Massive project maps DNA tags that define each cell's identity. *Science Magaine News* .

Poundston, W. 2005. *Fortune's Formula: The Untold Story of the Scientific Betting System That Beat the Casinos and Wall Street.* Hill and Wang.

Rands, C.M., Meader, S., Ponting, C.P., and Lunter, G. 2015. 8.2% of the human genome is constrained: Variation in rates of turnover across functional element classes in the human lineage. *PLOS ONE* 10(7). http://dx.doi.org/10.1371/journal.pgen.1004525

Rehman, J. 2013. The ENCODE controversy and professionalism in science. *The Next Regeneration Blog* Featured by Spektrum (scilogs.com) in association with Nature.com. http://www.scilogs.com/next_regeneration/the-encode-controversy-and-professionalism-in-science/

Saletore, Y., Meyer, K., Korlach, J., Vilfan, I.D., Jaffrey, S., and Mason, C.E. 2012. The birth of the epitranscriptome: deciphering the function of RNA modifications. *Genome Biology* 13.

Sancar, A. 2008. The intelligent clock and the Rube Goldberg clock. *Nature Structural & Molecular Biology* 13:23–24.

Sanford, J.C. 2014. *Genetic Entropy.* FMS Publications.

Satterlee, J. 2015. Enabling exploration of the eukaryotic epitranscriptome(e4). *Common Fund Epitranscriptomics Work Group* Division of program coordination, planning, and strategic initiatives of the National Institutes of Health. https://dpcpsi.nih.gov/sites/default/files/council%20jan%2030%202015%20Pres%20E4.pdf

Specified Complexity. 2016. *Wikipedia* Revision 751945651. https://en.wikipedia.org/wiki/Specified_complexity

Tale of the Peacock. 2001. *PBS Evolution Library* . http://www.pbs.org/wgbh/evolution/library/01/6/l_016_09.html

Tan, C.L. and Tomkins, J.P. 2015. Information processing differences between bacteria and eukarya—implications for the myth of eukaryogenesis. *Answers Research Journal* 8:143–162.

Thorp, E.O. 1961. A favorable strategy for Twenty-One. *Proceedings National Academy of Sciences* 47(1):110–112.

Thorp, E.O. 1966. *Beat the Dealer*. Vintage.

Thorp, E.O. 1967. *Beat the Market: A Scientific Stock Market System*. Random House.

Thorp, E.O. 1984. *The Mathematics of Gambling*. Gambling Times.

Trevors, J.T. and Abel, D.L. 2004. Chance and necessity do not explain the origin of life. *Cell Biology International* 28(11):729–739.

US Food and Drug Administration. 2015. FDA approves Farydak for treatment of multiple myeloma. *US Food and Drug Administration Website* . http://www.fda.gov/NewsEvents/Newsroom/PressAnnouncements/ ucm435296.htm

Zimmer, C. 2015. Is most of our DNA garbage? *New York Times* . http://www.nytimes.com/2015/03/08/magazine/is-most-of-our-dna-garbage.html?_r=0

14

Other Non-Naturalistic Methodologies in Modern Practice

JONATHAN BARTLETT AND ERIC HOLLOWAY

The Blyth Institute

Abstract

Some fields already incorporate alternatives to methodological naturalism. However, few people outside the field are familiar with these alternatives or how they are used. Sometimes these non-naturalistic methodologies are being used without the participants' cognizance that the methodology is not methodologically naturalistic. Here, we show a smattering of fields that we are aware of that have touched upon methodologies that don't depend on naturalism.

1 Introduction

While methodological naturalism has become the de facto standard in many fields, there have been a number of subfields of different disciplines that operate according to different rules. In some cases, the disconnect with methodological naturalism is not made explicit and, in fact, may be unknown to the participants. In other cases, the disconnect with methodological naturalism is clear and explicit.

The goals of this paper are:

1. to show that academic inquiry can be effective outside of naturalism

2. to promote non-naturalistic approaches from some fields to inspire similar techniques in other fields

3. to show how non-naturalistic thinking makes better sense of fields that are not aware of their use of non-naturalism, and how making their non-naturalistic aspects explicit can deepen their results

Hopefully, this paper provides inspiration and ideas for moving the non-naturalistic program forward in a variety of fields.

2   Methodological Dualism in Austrian Economics

One of the places where methodological naturalism has had an explicit challenge is in economics. In the late 1800s, a dispute arose within economics as to the role of history and the role of individuals. Menger (1883) criticized the so-called "historical" school of thought that considered economic activity entirely the result of preceding history. It could be viewed simply as a necessary outgrowth of what came before. Menger, instead, believed that it was the circumstances and choices of individuals (both as individuals and corporately) that drove the economy. According to Menger, the so-called "historical" school did not understand history and was of the opinion that while history had parallels and worked *within* useful theoretical systems, the historical school's simplistic view over-emphasized the degree to which corporate economic patterns were similar.

Menger noted that various parts of society could be separated into things that were comparable to *organisms* and things that were comparable to *mechanisms*. Those comparable to organisms were essentially ineffable to the science of economics, but those comparable to mechanisms were amenable to calculation. Therefore, humans are able to use their will to achieve ends by making mechanisms, and groups of humans are able to use their collective will to achieve ends by making social mechanisms. While economics can properly measure the effects and results of the *mechanisms* that are established by these acts of will, economics does not have access to the originating choices that instantiated them.

This idea of a separation of will and mechanism was more formally defined in von Mises (1949). Here, Mises defines what he calls *methodological dualism*. In this understanding, human choices are irreducible to physical phenomena and must be considered as first-class entities. Furthermore, even if human choices *could* be reduced to other phenomena, in our present state of understanding, we do not have access to this, and therefore, as a matter of *methodology*, choice must be considered an irreducible entity. As Mises explains:

> Concrete value judgments and definite human actions are not open to further analysis. We may fairly assume or believe that they are absolutely dependent upon and conditioned by their causes. But as long as we do not know how external facts—physical and physiological—produce in a human mind definite thoughts and volitions resulting in concrete acts, we have to face an insurmountable *methodological dualism*. In the present state of our knowledge the fundamental statements of positivism, monism and panphysicalism are mere metaphysical postulates devoid of any scientific foundation and both meaningless and useless for scientific research.

Reason and experience show us two separate realms: the external world of physical, chemical, and physiological phenomena and the internal world of thought, feeling, valuation, and purposeful action. No bridge connects—as far as we can see today—these two spheres. Identical external events result sometimes in different human responses, and different external events produce sometimes the same human response. We do not know why.

In the face of this state of affairs we cannot help withholding judgment on the essential statements of monism and materialism. We may or may not believe that the natural sciences will succeed one day in explaining the production of definite ideas, judgments of value, and actions in the same way in which they explain the production of a chemical compound as the necessary and unavoidable outcome of a certain combination of elements. In the meantime we are bound to acquiesce in a methodological dualism.

(von Mises, 1949, pgs. 17–18)

Thus, while methodological naturalism presumes that all events have causes that are reducible to mechanisms, methodological dualism takes as a methodological presumption the idea that human will is not reducible in this way. Similar to methodological naturalism, this does not directly *impose* the metaphysical viewpoint of dualism on the practitioner, but merely adopts it as a methodological requirement. However, such methodology certainly sits more comfortably with those who have a matching metaphysic.

This viewpoint has been extended in recent years through the work of Gilder (2013) and Thiel and Masters (2014). Gilder (2013) extends methodological dualism by adding a new component—human *creativity*. While traditional Austrian economics focuses on human *choices*, Gilder focuses on the creativity that is required for macroeconomic growth. According to Gilder, typical macroeconomic models that leave out human creativity and try to reduce the operation of the economy to an equation miss the mark entirely. The economy is not a grand equation such that if we put in the right values in the right places we will get economic growth. Bagus (2016) shows that perceiving the economy in this (naturalistic) way is what has led to the economic catastrophes of the twentieth and twenty-first centuries.

Instead, Gilder (2013) shows that the economy grows when individual creativity, which is not reducible to equations, is allowed to flourish and is given proper feedback. Therefore, the goal of economics is not to explicitly grow the economy (since we cannot, in principle, predict creativity), but rather to enable the conditions that allow creativity to operate most effectively—provide enough freedom, stability, and feedback to allow the creative operation to work. In other words, any attempt to dictate the performance of the economy will be fundamentally flawed. Instead, one needs to prepare the economy to better nurture and incorporate the unpredictable creativity of the economy's members.

Thus, this view of economics relies on a distinction between what can be known through mechanism (i.e., equations) and what requires non-mechanical input (i.e., creativity). Only by recognizing this distinction can economics properly aid in the growth of the economy.

Thiel and Masters (2014) provides a further expansion of this idea into microeconomics. In his book, Thiel recognizes the difference between what can be done on an individual basis via human cognition and what can be done via algorithm (i.e., equation). Thiel states:

> computers are far more different from people than any two people are different from each other; men and machines are good at fundamentally different things. People have intentionality—we form plans and make decisions in complicated situations. We're less good at making sense of enormous amounts of data. Computers are exactly the opposite: they excel at efficient data processing, but they struggle to make basic judgments that would be simple for any human....
>
> In 2012, one of [Google's] supercomputers made headlines when, after scanning 10 million thumbnails of YouTube videos, it learned to identify a cat with 75% accuracy. That seems impressive—until you remember that an average four-year-old can do it flawlessly. When a cheap laptop beats the smartest mathematicians at some tasks but even a supercomputer with 16,000 CPUs can't beat a child at others, you can tell that humans and computers are not just more or less powerful than each other—they are categorically different. (Thiel and Masters, 2014, pgs. 143–144)

Thiel uses this categorical difference between humans and computers to recognize that one of the most powerful ways to achieve high profits is to identify axioms that other people aren't aware of, which he illustrates in his question, "What important truth do very few people agree with you on?" (Thiel and Masters, 2014, pg. 12).

According to Thiel, developing new axioms is what leads to economic growth. Businesses based on existing knowledge and ideas move the economy from 1 to N, but an axiom can move the economy from 0 to 1. In other words, new axioms allow the creation of completely new areas for economic development. Once the axiom is developed, it can then be replicated and used by others throughout the economy, but the primary *driver* for growth is attaining the axiom in the first place. This idea was also indicated in theoretical studies in the *Engineering and Metaphysics* conference (Bartlett, 2014b,a).

As indicated by Robertson (1999) and Bartlett (2014b), the development of new axioms cannot be understood in a naturalistic way. Thus, we can see that in economics non-naturalistic methodologies allow us to better recognize the types of phenomena we are seeing as well as better understand their relationships.

3   Human Computation and Artificial Artificial Intelligence

The commitment to methodology is a double edged sword. On the one hand, it can enforce a point of view that ignores facts, but on the other hand, a fact-based methodology can enforce a point of view that is contrary to the reigning paradigm. We have seen this in economics, where necessity forces economists to adopt a dualistic perspective regarding the human and mechanical realms of the economy.

Similarly, computer science is faced with the necessity of methodological dualism. In the age of widespread social technologies, such as the personal computer, the Internet, and Facebook, humans are both the consumer and the product. Computer scientists will give lip service to the monism of computer science, General Artificial Intelligence (GAI), but agree that until GAI arrives it is much more useful to include the human in the loop. There is great disillusionment with the promise of GAI because it has been 'imminent' since Alan Turing invented the Turing machine more than half a century ago.

Methodological dualism in computer science is particularly apparent in the rise of Human Computation (HC) and Artificial Artificial Intelligence (AAI) (The Economist, 2006). Human computation is the use of human generated solutions to tasks for which there is no known algorithmic solution. The tasks are normally micro-tasks, which can be completed quickly by most Internet users with minimal-to-no training. Examples of these tasks are identifying objects in images, identifying parts of speech in sentences, and transcribing audio recordings. These tasks are trivial for a child but elude the most powerful supercomputers.

The micro-tasks are aggregated algorithmically and/or using further human input (Dai, Mausam, and Weld, 2011), and the resulting hybrid system is termed Artificial Artificial Intelligence. The system gives the appearance of algorithmically solving a particularly problem by looking like traditional artificial intelligence, but the inner workings are fundamentally dependent upon continuous human interaction.

The concept of human computation first caught the public's attention through the pioneering work of Luis Von Ahn's now ubiquitous CAPTCHAs (Law and von Ahn, 2011) and the scientific breakthroughs of the Foldit project. One of the most well-known breakthroughs made headlines because amateurs reverse engineered the crystalline structure of an HIV protease, a feat that has escaped the most powerful supercomputers and experts (Khatib, DiMaio, Cooper, Kazmierczyk, Gilski, Krzywda, Zabranska, Pichova, Thompson, Popović, Jaskolski, and Baker, 2011).

Capitalizing upon these successes, Amazon has released a public micro-task platform, Mechanical Turk (named after "The Turk," an eighteenth century HC originator), that is widely used by academic researchers and Internet companies (Bolt, 2005). However, human computation is not merely a niche interest. Companies such as Google, Facebook, and Microsoft are so reliant on HC to make their algorithms run that they have created their own internal platforms (Marcus and Parameswaran, 2015). Arguably, HC is the fuel powering the entire Internet revolution. It is human-produced information that is so widely sought after on all Internet platforms.

In retrospect, it is ironic that artificial intelligence has become even more hyped as the necessity of methodological dualism becomes ever more pronounced. AI is the corollary of methodological naturalism in computer science since AI is predicated upon reducing the mind to hardware. Yet, as a quote widely attributed to Yogi Berra has told us, "In theory there is no difference between theory and practice. In practice there is."[1] The need for practitioners to discard methodological naturalism in favor of dualism to get things done cautions us that methodological open-mindedness is more pragmatic than methodological dogmatism.

4   Moral Philosophy in Online Transaction Analysis

One of the main pitfalls in online commerce platforms is the issue of detecting fraudulent charges. The difference between whether an online enterprise is profitable or unprofitable often rests on the platform's ability to prevent or catch fraudulent transactions. The Internet creates an especially problematic source for fraud as there is no distance on the Internet between the person perpetrating the fraud and the company being defrauded. While in the physical world there is a limit to the how much fraud can occur in a given locality based on the number of people willing to commit fraud, on the Internet there is not a similar natural barrier. Because of this, fraud detection and prevention have become very problematic areas for high-profile commercial websites.

The book *Start-Up Nation* Senor and Singer (2011) recounted the story of how an enterprising company, Fraud Sciences, used moral philosophy to quickly and accurately detect fraudulent transactions on PayPal. It describes the meeting between Fraud Sciences' Shvat Shaked and PayPal's Scott Thompson, as going something like this:

> "So what's your model, Shvat?" Thompson asked, eager to get the meeting over with. Shifting around a bit like someone who hadn't quite perfected his one-minute "elevator pitch," Shaked began quietly: "Our idea is simple. We believe that the world is divided between good people and bad people, and the trick to beating fraud is to distinguish between them on the Web."

---

[1] This quote was first attributed to an anonymous source in Savitch (1984) (overheard at a computer science conference) and was later attributed to both Yogi Berra (improbable) and Jan L. A. van de Snepscheut (more likely).

Thompson suppressed his frustration. This was too much, even as a favor to Benchmark. Before PayPal, Thompson had been a top executive at credit card giant Visa, an even bigger company that was no less obsessed with combating fraud. A large part of the team at most credit card companies and online vendors is devoted to vetting new customers and fighting fraud and identity theft, because that's where profit margins can be largely determined and where customer trust is built or lost.

Visa and the banks it partnered with together had tens of thousands of people working to beat fraud. PayPal had two thousand, including some fifty of their best PhD engineers, trying to stay ahead of the crooks. And this kid was talking about "good guys and bad guys," as if he were the first to discover the problem.

"Sounds good," Thompson said, not without restraint. "How do you do that?"

"Good people leave traces of themselves on the Internet—digital footprints—because they have nothing to hide," Shvat continued in his accented English. "Bad people don't, because they try to hide themselves. All we do is look for footprints. If you can find them, you can minimize risk to an acceptable level and underwrite it. It is really that simple.

Thompson was beginning to think that this guy with the strange name had flown in not from a different country but rather a different planet. Didn't he know that fighting fraud is a painstaking process of checking backgrounds, wading through credit histories, building sophisticated algorithms to determine trustworthiness? You wouldn't walk into NASA and say, "Why build all those fancy spaceships when all you need is a slingshot?"

(Senor and Singer, 2011, pgs. 24–25)

In the end, the Fraud Sciences model performed both faster and more accurately than PayPal's own system using less data. Fraud Sciences was able to score 17% better on PayPal's most troubling category—good customers that are mistakenly flagged as bad.

This shocked PayPal executives precisely because PayPal had the most advanced system for fraud checking in the world, yet their data scientists were easily defeated by a no-name startup talking about "good guys" and "bad guys."

At the end of the day, what enabled Fraud Sciences to better analyze transactions was not their ability to make *computer models* of fraud, but rather to engage in *moral philosophy* to determine which pieces of data they needed. Because they were able to recognize an important fact about evil—that it likes to hide in the dark—they were able to understand what all the other models missed. That is, if someone is living their life in the full light of day, it is unlikely that they are engaging in fraud.

Large-scale businesses have no choice but to operate based on data and algorithms. However, the story of Fraud Sciences shows that approaching problems with a philosophy-first attitude can often yield quantitative benefits.

5   Engineering Principles in Systems Biology

In the nineteenth and twentieth centuries, biologists increasingly began studying organisms using naturalistic methodologies. Since naturalism connects all events through a historical framework, biology itself started to become entirely enmeshed in the historical framework of evolution. Likewise, since naturalism breaks all events down into their constituent components, this was done in biology as well.

Thus, in the nineteenth and twentieth centuries, biology was characterized by reference to evolutionary history and physico-chemical reductionism. The former attempted to understand every part of an organism in terms of the historical accidents and selective constraints that happened within the ancestry of the organism. The latter attempted to understand the organism's every action in terms of smaller actions on lower and lower levels.

I do not mean to say here that these were necessarily problematic. As a matter of fact, the nineteenth and twentieth centuries can boast of great advances along both of these lines. The problem, as always, stems from looking at only certain types of causes and ignoring others. As the saying goes, "If all you have is a hammer, everything starts to look like a nail." It is the result of taking Occam's Razor to extremes where it is not appropriate to do so.

Carl Woese has reflected on the problems that this mode of biology has presented:

> Molecular biology's success over the last century has come solely from looking at certain ones of the problems biology poses (the gene and the nature of the cell) and looking at them from a purely reductionist point of view. It has produced an astounding harvest. The other problems, evolution and the nature of biological form, molecular biology chose to ignore, either failing outright to recognize them or dismissing them as inconsequential, as historical accidents, fundamentally inexplicable and irrelevant to our understanding of biology. Now, this should be cause for pause.

(Woese, 2004, pg. 175)

Woese comes from a position of methodological naturalism, and he is not proposing any deviation from this. Nonetheless, it is interesting that the timeframe that most missed out on the big questions of biology in exchange for the little ones was also the era most dominated by methodological naturalism.

The new science of *systems biology* is an attempt to pursue biology without the constraints of historicism and reductionist mechanisms. In systems biology, biological systems are analyzed as holistic units without heavy regard for their history. Instead, systems biologists seek the top-level design principles of a biological system. Systems biology analyzes biology at multiple scales showing how the patterns at one scale interact with patterns at another scale.

Unlike naturalism, which privileges the lowest-scale causes and their mechanics, systems biology privileges the design principles by which systems operate. Naturalism, likewise, favors explanations of a system in terms of historical causes, such evolving through previous systems, while systems biology favors holistic explanations focused on the purposes of the systems in question.

As such, while systems biology does not itself explicitly exclude methodological naturalism, it seems that a non-naturalist perspective makes the study of systems biology more cohesive and understandable. Physics does not have a category of "design principles," so if a design principle is found, in what sense could it be a part of naturalism? However, if design is considered a causative principle in and of itself (contra naturalism), then it makes much more sense of what is happening in systems biology. As Nelson (2016) points out, in designed systems, knowing the purpose of the design bears directly on the question of how it is designed to work.

An additional principle noted by Nelson (2016) is what he calls the *inference from system-level functional necessity*. Oftentimes, biologists are faced with systems where the operation is only partially known. There are two ways that inferences are generally made about the unknown parts of biological systems—historical and system-level functional necessity. Nelson points out that, on the historical viewpoint, biologists frequently make assumptions about the operation of systems that stem from their knowledge of the natural history of the system and the kinds of changes that natural selection (or other evolutionary mechanisms) are likely to impart to a given biological system. On the other hand, the inference from system-level functional necessity uses the top-level design *requirements* of the system as a guide to the parts that are unknown. In other words, if we imagine that we are given a top-level description of the biological system, we can often infer the parts we don't know from that description. This method of inference privileges design above historicism and bottom-up mechanism, the pillars of naturalism. What Nelson shows is that, when the inferences from history and system-level functional necessity are in conflict, the inference from system-level functional necessity has almost always been shown to be correct in the long run.

In an even more explicit end-run around naturalism, Halsmer, Gewecke, Gewecke, Roman, Todd, and Fitzgerald (2014) shows that the biological sciences, when they are most successful, perform the same type of work that engineers do when reverse-engineering machines built by other designers. Halsmer suggests that making this role of reverse-engineer more explicit in biology—by explicitly studying and adopting the reverse-engineering literature from engineering sciences—would serve to improve the ways that biologists perform their tasks.

6   Conclusion

As can be seen, many areas of inquiry can benefit and are already benefitting from looking beyond methodological naturalism. This is usually taken in the form of recognizing non-physical phenomena explicitly as a first-class, causally real principle. In economics and computer science, it is by establishing human choices and creativity as fundamental causal principles. In online transaction analysis, it is recognizing that moral philosophical categories are prior and more fundamental to the bits of data being collected. In biology, it is recognizing design principles as being more fundamental than the history or the physical reductions themselves.

In some cases the recognition that this type of study goes beyond methodological naturalism is explicit, and in some cases the break with methodological naturalism, while real, is not yet recognized or understood. Hopefully, these examples can serve as templates for further exploration into non-naturalistic methodologies.

References

Bagus, P. 2016. Methodological naturalism in the austrian school of economics. *2016 Conference on Alternatives to Methodological Naturalism* .

Bartlett, J. 2014a. Measuring software complexity using the halting problem. In *Engineering and the Ultimate*, pp. 123–130, Blyth Institute Press, Broken Arrow, OK.

Bartlett, J. 2014b. Using Turing oracles in cognitive models of problem-solving. In *Engineering and the Ultimate*, pp. 99–122, Blyth Institute Press, Broken Arrow, OK.

Bolt, K.M. 2005. Amazon creates artificial artificial intelligence. *Seattle Pi* . http://www.seattlepi.com/business/article/Amazon-creates-artificial-artificial-intelligence-1186698.php

Dai, P., Mausam, and Weld, D.S. 2011. Artificial intelligence for artificial artificial intelligence. *Proceedings of the Twenty-Fifth AAAI Conference on Artificial Intelligence* . https://www.aaai.org/ocs/index.php/AAAI/AAAI11/paper/viewFile/3775/4051

Gilder, G. 2013. *Knowledge and Power: The Information Theory of Capitalism and How it is Revolutionizing our World.* Regnery Publishing.

Halsmer, D., Gewecke, M., Gewecke, R., Roman, N., Todd, T., and Fitzgerald, J. 2014. Reversible universe: Implications of affordance-based reverse engineering of complex natural systems. In *Engineering and the Ultimate*, pp. 11–38, Blyth Institute Press, Broken Arrow, OK.

Khatib, F., DiMaio, F., Cooper, S., Kazmierczyk, M., Gilski, M., Krzywda, S., Zabranska, H., Pichova, I., Thompson, J., Popović, Z., Jaskolski, M., and Baker, D. 2011. Crystal structure of a monomeric retroviral protease solved by protein folding game players. *Nature Structural & Molecular Biology* 18(10):1175–1177.

Law, E. and von Ahn, L. 2011. *Human Computation.* Morgan and Claypool.

Marcus, A. and Parameswaran, A. 2015. Crowdsourced data management: Industry and academic perspectives. *Foundations and Trends in Databases* 6(1-2):1–161.

Menger, C. 1883. *Investigations into the Method of the Social Sciences with Special Reference to Economics.*

Nelson, P. 2016. Design triangulation. In *2016 Conference on Alternatives to Methodological Naturalism*, video.
https://www.youtube.com/watch?v=rNY_i1kJAnk

Robertson, D.S. 1999. Algorithmic information theory, free will, and the turing test. *Complexity* 4(3):17–34.

Savitch, W.J. 1984. *PASCAL: An Introduction to the Art and Science of Programming*. Benjamin-Cummings Publishing Company.

Senor, D. and Singer, S. 2011. *Start-up Nation: The Story of Israel's Economic Miracle*. Twelve.

The Economist. 2006. Artificial artificial intelligence .
http://www.economist.com/node/7001738

Thiel, P. and Masters, B. 2014. *Zero to One: Notes on Startups, or How to Build the Future*. Crown Business.

von Mises, L. 1949. *Human Action: A Treatise on Economics*. Yale University Press.

Woese, C.R. 2004. A new biology for a new century. *Microbiology and Molecular Biology Reviews* 68(2):173–186.

About the Authors

Editors

Jonathan Bartlett

Jonathan Bartlett is the Director of The Blyth Institute, a non-profit research and education organization focusing on pioneering new non-reductionistic approaches to biology. Jonathan's research focuses on the origin of novelty—both the origin of biological novelty in adaptation as well as the origin of insight in the human creative process. Jonathan is the author of several textbooks, and was also the lead editor for the 2014 book *Engineering and the Ultimate: An Interdisciplinary Investigation of Order and Design in Nature and Craft*.

Eric Holloway

Eric Holloway recently joined The Blyth Institute as a research fellow. Eric's research focuses on new directions in artificial intelligence and the development of the field of human computation. His research also includes the use of evolutionary algorithms and swarm intelligence in cybersecurity. Eric's research is a rare mix of the theoretical and the practical and has attracted attention and funding from groups as diverse as the Center for Evolutionary Informatics and the Air Force Research Lab.

Chapter Authors

Salvador Cordova

Salvador Cordova holds a Master of Science degree in Applied Physics from Johns Hopkins University, a BS in Mathematics with a minor in Physics, a Bachelor of Science degree in Electrical Engineering with a minor in music, and a Bachelor of Science degree in Computer Science. In addition to managing a small hedge fund, he is also working part-time as a research assistant in the field of bioinformatics as well as preparing for doctoral studies in molecular biophysics. He has been featured on national TV, books, magazines, radio shows, newspapers, and science journals in his many different endeavors.

Tom Gilson

Tom Gilson has a master's degree in industrial and organizational psychology. He has served as the National Field Director and Vice President for Strategic Services with Ratio Christi, and is currently senior editor for the web magazine *The Stream*. His Blog, ThinkingChristian.net, is one of the top Christian philosophy blogs on the web.

James D. Johansen

James D. Johansen (MSEE, MASR, MACA) has a variety of educational and professional experiences that equip him for his interests of bridging scientific and theological themes. He has Bachelor and Master of Science degrees in Electrical Engineering from the University of Southern California. Working for several research and development companies he has been active in leading and managing technology development and application supporting a variety of organizations like NASA and NSF, along with academic and industry collaboration. Having a desire to more effectively explore the integration of science and theology, he studied at Biola University and earned a Master of Arts in Science and Religion and a Master of Arts in Christian Apologetics. At Biola he looked at ways to examine genomic information by integrating scientific, theological, and philosophical perspectives. On completion of this work, he wanted to explore these themes more deeply and is working on a PhD at Liberty University in Theology and Apologetics with an emphasis on the philosophy of science. Jim is active in a variety of professional organizations, including IEEE, ASA, and ETS, which includes presenting at various conferences both domestically and internationally. He is active at his church where he teaches and leads small groups that leverage his Biblical and theological training.

James C. LeMaster

Jim LeMaster, a native of Seattle, holds four degrees, including two from the University of Washington, one from a seminary in Taiwan, and most recently a PhD in Applied Apologetics. Jim and his family served as missionaries in East Asia for 14 years. Since 2009 he has been carrying out what he calls an outreach and persuasion ministry to international students and scholars. Currently Jim has been doing this work at the University of Kentucky and the University of Louisville. During the last ten years, he has taught courses, given presentations, and interacted extensively with students and scholars—both in the U.S. and in Asia—about the topics of worldviews, philosophy, evidences for design from the natural sciences, and other topics. Jim currently lives in Louisville, Kentucky with his wife Ruth. They have three children.

Mario López

Mario A. López is the president and founder of the Organización Internacional para el Avance Científico del Diseño Inteligente (OIACDI). Mario has been actively involved with the Intelligent Design (ID) movement since the mid-1990s and has served on the board of directors of various ID organizations. In his current project, OIACDI, he gathers scientists and scholars from Spanish-speaking countries to collaborate in bringing ID to their communities. Working with a forensics geneticist, an informatics engineer, a pathologist, a materials scientist, a mathematician, and a biochemist, he and his colleagues launched ID beyond our borders through the publication of several books, including: *Elementos de Estructuras Funcionales*, *La Quinta Vía Y El Diseño Inteligente*, and *Neotomismo, Mecanicismo Y Diseño Inteligente*. In addition, they took up the task of translating material from leading ID proponents and making it accessible to Spanish readers around the globe. Mario has co-authored a book entitled *Charles Darwin Frente Al Diseño Inteligente* and has written several online articles that attempt to disambiguate intelligent design theory for the general reader. His ideas on the demarcation problem will be further developed in his forthcoming book, *Beyond the Pale of Science*.

Arminius Mignea

Arminius Mignea received his Engineering Diploma from the Polytechnic University of Bucharest, School of Computers. He worked as a software engineer and researcher at the Institute for Computing Technique for 14 years. Employed in the operating systems laboratory, he wrote system software in macro assembly language, Pascal, and C, and published some twelve research papers alone or as part of the team. After immigrating to the United States and settling in California's Silicon Valley, he worked for numerous technology startups developing both server and front-end software for network, enterprise, and security management products. More recently he was involved in the specification, architecture design, and development of software systems for enterprise network traffic monitoring, software build management, and TV server systems. About ten years ago, Arminius became interested in the designs, architectures, and machinery of biological systems. Arminius is fascinated with the beauty, intricacies, coordination, and incredible levels of organization in living things and marvels at the skills of their architect.

Sam S. Rakover

Sam S. Rakover is a professor emeritus at the department of psychology, Haifa university, Haifa, Israel. His areas of expertise are face perception and recognition, philosophy of science, and the mind. He published four professional books and about one hundred papers. His new book *Explaining Behavior: A New Methodological Approach* will be published by Lexington. He is the author of ten novels, many short stories, and three board games.

Noel Rude

Noel Rude is the author of many articles on linguistics in the *International Journal of American Linguistics* and has contributed chapters to several edited volumes, including *External Possession* and *University of British Columbia Working Papers in Linguistics*. His *Umatilla Dictionary* documents the language of the Umatilla people, who are east of the Cascade Mountains in Oregon and Washington. Working for many years with the accumulated scholarship of linguists and anthropologists as well as with elders on the Umatilla Reservation, tribal linguist Noel Rude has painstakingly recorded and rationalized words, pronunciations, phrases, and other elements of the Umatilla language.

About the Blyth Institute

The Blyth Institute is a non-profit research and education organization focusing on non-reductionist approaches to biology specifically and science in general. The Blyth Institute's biology research includes studying patterns of mutation in the genome, studying patterns of variation across taxonomic branches, and analyzing molecular behavior of cells in terms of design patterns in computer engineering. The Blyth Institute also studies human cognition, focusing on creativity in cognitive models. Finally, The Blyth Institute studies scientific methodologies with the aim to find better methodologies that capture and analyze phenomena not previously amenable to rigorous analysis. The Blyth Institute provides grants to scholars working in related areas.

The Blyth Institute also engages in a number of educational and outreach activities. These activities include:

- Guest lecturing in universities about the research done by the Institute

- Giving public lectures on recent advances in biology and related fields

- Educating high-school students on a variety of topics in mathematics and biology

- Building and operating a mobile microscopy lab program for introducing high-school students to microscope work

- Running a microscope leasing program to allow independent and homeschool classrooms to have access to better microscopy equipment than they could otherwise afford

The Blyth Institute is an independent organization which operates entirely on private funding. You can find out more information about The Blyth Institute on the web, at `http://www.blythinstitute.org/`. If you would like to contact The Blyth Institute on any of these topics, please email `info@blythinstitute.org`.

About the Center for Evolutionary Informatics

The Center for Evolutionary Informatics explores the conceptual foundations, mathematical development, and empirical application of evolutionary informatics. Evolutionary informatics merges theories of evolution and information, thereby wedding the natural, engineering, and mathematical sciences. Evolutionary informatics studies how evolving systems incorporate, transform, and export information. The principal theme of the center's research is teasing apart the respective roles of internally generated and externally applied information in the performance of evolutionary systems.

About the Conference

The *2016 Conference on Alternatives to Methodological Naturalism* is the first in a series of conferences on developing methodological alternatives to naturalism in academic research. While many individuals, organizations, and conferences have discussed whether or not methodological naturalism is a necessary rule of science, most of them stay in the realm of philosophy.

The goal of the *Alternatives to Methodological Naturalism* series of conferences is to move toward the practical side and describe *how* science can be accomplished in non-naturalistic contexts. Specifically, what methodological features need to be modified, and what are the implications of those changes to the content and reliability of scientific work.

As is evident from the many fields represented in the present volume, questions about naturalism extend to a variety of sciences, including biology, psychology, economics, and computer science. Our goal is to create a multidisciplinary conversation about how naturalism has affected these areas for good or for bad, and outline what sort of methodologies can be changed to allow for a wider picture of reality. Many disagree that any change needs to be made, and we welcome their voices, too.

For more information about the *Alternatives to Methodological Naturalism* series of conferences, visit the website at `http://www.am-nat.org/` and sign up for the newsletter to receive up-to-date information on our conference and publication schedule.

Index

www.ingramcontent.com/pod-product-compliance
Lightning Source LLC
Chambersburg PA
CBHW050402110426
42812CB00006BA/1780